Springer Series in Optical Sciences Volume 44

Springer Series in Optical Sciences

Volume 42 **Principles of Phase Conjugation**
By B. Ya. Zel'dovich, N. F. Pilipetsky, and V. V. Shkunov

Volume 43 **X-Ray Microscopy**
Editors: G. Schmahl and D. Rudolph

Volume 44 **Introduction to Laser Physics**
By K. Shimoda

Volume 45 **Scanning Electron Microscopy**
By L. Reimer

Volume 46 **Holography in Deformation Analysis**
By W. Schumann, J.-P. Zürcher, and D. Cuche

Volumes 1-41 are listed on the back inside cover

Koichi Shimoda

Introduction to Laser Physics

With 87 Figures

Springer-Verlag
Berlin Heidelberg New York Tokyo 1984

Professor Koichi Shimoda
Faculty of Science and Technology, Keio University, 3-14-1 Hiyoshi, Kohokuku,
Yokohama 223, Japan

Revised translation of the original Japanese edition:
Koichi Shimoda: Rêzâ Butsuri Nyûmon

Originally published in Japanese by Iwanami Shoten, Publishers, Tokyo (1983)
English translation by Munetada Yamamuro

ISBN 3-540-13430-1 Springer-Verlag Berlin Heidelberg New York Tokyo
ISBN 0-387-13430-1 Springer-Verlag New York Heidelberg Berlin Tokyo

Library of Congress Cataloging in Publication Data. Shimoda, Kōichi. Introduction to laser physics. (Springer series in optical sciences ; v. 44) Rev. translation of: Koichi Shimoda: Rêzâ Butsuri Nyûmon. 1. Lasers. I. Title. II. Series. QC688.S55 1984 535.5'8 84-5629

Typesetting: Schwetzinger Verlagsdruckerei, 6830 Schwetzingen
Offset printing: Beltz Offsetdruck, 6944 Hemsbach/Bergstr. Bookbinding: J. Schäffer OHG, 6718 Grünstadt.
2153/3130-543210

Preface

The laser has opened a new field of science and technology and its progress is making great strides. Various types of lasers have developed to such an extent that not only are they useful in industrial applications, they have also brought about such innovations as the laser processing of microstructures of semiconductor and other materials, the state selection of atoms and molecules, single-atom detection and trapping, femtosecond measurements, phase conjugation, and many others.

Accordingly, those who are engaged in laser engineering are often required to understand and pursue the physical principles of the laser; this goes without saying for scientists studying laser spectroscopy and quantum optics.

Now, the purpose of this book is to provide an introductory course of comprehensive laser physics for students with a background of electromagnetic theory, basic quantum mechanics, and calculus. The text will afford an effective understanding of the basic concepts, principles, and theoretical treatments of the laser as well as laser-induced effects. However, laser physics is still under development and it has not yet been fully established in any accepted structure. The approach presented in this book is by an author who has been involved in a substantial part of laser research from the time before the advent of the laser to the present.

Most of the mathematical treatments and physical ideas above an elementary level are explained in detail. The subject matter is treated with full physical interpretation providing the reader with a concrete foundation in laser physics. Discussing the orders of magnitude of physical quantities and practical parameters is stressed, rather than theoretically calculating exact values.

The first chapter is devoted to an overview of the state of the art of most typical lasers. Modern optics relevant to lasers is described in the subsequent chapters followed by the physical processes of emission and absorption of light. Then, the principle of the laser and the concept of population inversion are described and a rate-equation approach is used to present the output characteristics of the laser. Atomic coherence and nonlinear polarization are described in detail to provide the semiclassical theory of lasers, nonlinear optics, and laser spectroscopy. Fully quantum-mechanical theories of the laser are reviewed briefly. Each chapter is accompanied by problems, most of which are not mere numerical calculations, but contain instructive and informative material. Hints and short answers are given.

The author is very grateful to Prof. Munetada Yamamuro who translated the Japanese edition. Without his efforts, this revised version in English would not have been realized. Particular thanks should go to Dr. Helmut Lotsch and Ms. Gillian Hayes of Springer-Verlag for their careful reading and correction of the manuscript. They have furthermore supplied additional reading and some references for the manuscript, and encouraged the author to prepare the problems. Finally he thanks Mrs. Chitose Hikami for her excellent typing of the manuscript and his students for their examination of many derivations in the text and answers to the problems.

Yokohama, Japan, April, 1984 *Koichi Shimoda*

Contents

1. The Laser – An Unprecedented Light Source

The word LASER is an acronym for "Light Amplification by Stimulated Emission of Radiation". The principle of the laser is the same as that of the MASER (Microwave Amplification by Stimulated Emission of Radiation) which was invented in 1954. Around the time when the laser was invented in 1960, it was called an optical maser or an infrared maser. The word laser has been generally accepted since about 1965.

The laser brought about a revolution in optical technology and spectroscopy, and had a far-reaching influence in various fields of science and technology. In the life sciences, medical sience, and even in nuclear fusion, many research projects are being carried out using lasers. Furthermore, there are natural phenomena such as self-focusing of light and optical bistability which could only have been discovered through the use of lasers.

Before describing the principles of the laser and investigating its characteristic behavior we shall, in this chapter, survey what kind of light laser radiation is, and what are the different types of laser.

1.1 The Properties of Laser Light

It may be said that the laser is an oscillator or an amplifier of light waves. Some lasers generate visible light but others generate infrared or ultraviolet rays which are invisible. Although the wavelength of visible light extends from about 0.37 to 0.75 μm, lasers with wavelengths between 0.1 μm in the vacuum ultraviolet and 1 mm in the microwave region are known. Those of practical use have wavelengths in the range 0.2 to 500 μm (0.5 mm). There are various types of lasers, ranging in size from semiconductor lasers, which are smaller than 1 mm, to lasers used in nuclear fusion experiments, which can be as large as 100 m. However, their basic properties are more or less the same.

1.1.1 Directivity

On watching visible laser light, it is at once recognizable that it is a narrow beam propagating in almost a straight line, unless it is reflected or refracted. However, even the laser radiation is not a perfectly parallel flux of light, but at large distances it gradually broadens due to diffraction, which can be explained by wave optics or the electromagnetic theory of light. The directivity of the beam is expressed by the angular broadening $\delta\theta$, which is related to the beam diameter d at the exit of the laser and its wavelength λ by the equation

$$\delta\theta \simeq \frac{\lambda}{d} . \tag{1.1}$$

The distribution of field intensity within the cross-section of the laser beam and the angular distribution at large distances are related by a Fourier transform as is well known in the theory of diffraction. Equation (1.1) is an approximate expression for the angle of directivity for a relatively smooth intensity distribution (Sect. 3.7). It is seen that for a red laser beam having a diameter of 2 mm and a wavelength of 0.6 μm, $\delta\theta$ is about 3×10^{-4} rad and the laser beam at a distance of 100 m, for example, broadens to only about 3 cm in diameter.

The reason why the directivity of laser radiation is so high is because the phase of the light waves over the cross-section of a laser beam is almost fixed everywhere. Such uniformity in phase is generally known as coherence. The chief characteristic of the laser is, in short, that it generates coherent light. Not only is laser light coherent in space but it is also coherent in time, so as to be almost perfectly monochromatic. Coherence will be discussed in some detail in Chap. 2.

1.1.2 Monochromaticity

A single line in the spectrum of light generated in an ordinary discharge tube or a spectral lamp is usually said to be monochromatic. But if we examine it with a spectrometer of high resolution, the spectral line will be seen to have some width. However, the width in wavelengths of laser light is remarkably narrow and even the very best spectrometer would fail to detect any measurable width. Moreover, if two laser beams of almost the same wavelength are incident in parallel on a detector, a beat will be observed at a frequency corresponding to the difference of the two laser frequencies. It is possible, therefore, to detect fluctuations in the laser frequency from the observed fluctuations in beat frequency. The spectral width of most lasers is between 1 MHz and 1 GHz, although it is known to be less than 1 Hz for a stable gas laser. Even a spectral width of 1 GHz for a visible laser is only one part in two million of the laser frequency, and it is equivalent to a width in wavelength of 1 pm (0.01 Å). The spectral width of a stable laser of 1 Hz corresponds to a frequency purity of the order of 10^{-15}.

Unlike the light waves from an ordinary light source, which are emitted as a succession of irregular pulses, the oscillation of laser light is almost a pure sinusoidal wave continuing for quite a long time. This is proved from an experiment in which a single laser beam is divided into two and then brought together again. It is well known that the two superposed beams exhibit clear interference fringes even when they have a considerable path difference between them (Chap. 2). The theoretical limit to the fluctuation of the laser frequency is determined by the spontaneous emission (Sect. 4.3); if P is the power of the light emitted from the active medium by stimulated emission and ν the frequency, the fluctuation is given by [1.1]

$$\delta\nu = \frac{2\pi h\nu(\Delta\nu)^2}{P} , \tag{1.2}$$

where $h = 6.626 \times 10^{-34}$ J · s is the Planck constant and $\Delta\nu$ is the halfwidth of the optical resonator used in the laser at half the maximum resonant intensity.

In an actual laser the resonance frequency of the resonator is perturbed for various reasons. Moreover, the resonance frequency fluctuates because the conditions for exciting the active medium are not stable. Thus, in order to stabilize the laser frequency it is necessary that the resonator should be constructed to be free from external vibration, change in dimensions due to temperature variation, and fluctuations in pressure and electromagnetic fields.

1.1.3 Energy Density and Brightness

The efficiency of a laser is not high. The output power of an ordinary laser is below 0.1% of the input and even that of the most efficient laser is no more than 40%. The output power of the small He-Ne laser, which is commonly used in optical experiments and measurements, is about 1 mW. It is weaker than a miniature light bulb. Even the power of a large Ar ion laser or a YAG laser is only 10–100 W (Sects. 1.3 and 2, respectively), about the same as a fluorescent lamp. However, because of its good directivity the laser light can be focused to a diameter equal to only a few times the wavelength using a lens of short focal length. Consequently, the power density of the light reaches a very high value at the focus. For example, even a small output power of 1 mW focused to an area of 10 μm^2 produces a power density of 10 kW cm^{-2} = 100 MW m^{-2}, and a focused power of 100 W will give a value of 1 GW cm^{-2} = 10 TW m^{-2}. With such focusing of the laser power it is possible to produce a very strong electric field which can by no means be obtained by dc or low-frequency ac electric circuits. The optical electric field in a 100 W laser beam focused to 10 μm^2 is calculated to be about 6×10^7 V/m; if a pulsed laser with a peak power of 100 MW is focused, an extremely strong electric field of 6×10^{10} V/m, that is, 60 gigavolts per meter, can be obtained.

The high laser power density enables us to attain a high concentration in the number of photons. For example, a laser power of 1 mW focused to give a power density of 10 kW cm^{-2} corresponds to a flux of 3×10^{22} photons cm^{-2} per second at a wavelength of 0.6 μm or a photon energy of 3.3×10^{-19} J. This means a concentration of 10^{12} photons cm^{-3}, since the speed of light is $c = 3 \times 10^8$ m/s. If the power density is 1 GW cm^{-2}, we will obtain a very high density of 10^{17} photons cm^{-3}. Consequently, if we focus laser light onto matter, several photons will often react simultaneously with a single atom.

Since the monochromaticity of laser light is very high, the light energy is concentrated within a very narrow spectral linewidth. Thus, even though the absolute magnitude of the laser power is low, the brightness is very high. If P is the laser power and its spectral frequency width is $\delta\nu$, the equivalent brightness temperature T_B is given by

$$T_B = \frac{P}{k_B \delta\nu} ,$$

where $k_B = 1.38 \times 10^{-23}$ J/K is the Boltzmann constant. For example, if $P = 1$ mW and $\delta\nu = 1$ Hz, we will obtain the exceedingly high brightness temperature of about 10^{20} K. Even when the spectral linewidth is as wide as $\delta\nu = 1$ MHz, for a larger power of $P = 10$ W, T_B will be 10^{17} K. In any case, when we compare these temperatures with the brightness temperature of the sun or an incandescent lamp – about 10^4 K – they are many orders of magnitude higher.

1.1.4 Ultrashort Pulses

The laser may generate an output of nearly constant amplitude and phase, but it may also be modulated at a very high speed. In most cases, AM (amplitude modulation) and FM (frequency modulation) up to microwave frequencies are possible, which means that pulses shorter than 1 ns can be obtained. When a modulator is coupled externally with a laser, it is the bandwidth of the external modulator which determines the short-pulse limit, but when the modulator is inside the laser resonator, it is mainly the gain bandwidth of the active medium which determines the limit of the pulse duration [1.2].

In ordinary electronics 1 ns is about the limit on the pulse width or the response time. On the other hand, laser pulses can be obtained which are about 1000 times shorter, or 1 ps (picosecond = 10^{-12} second), and, more recently, subpicosecond pulses of 0.1–1 ps and shorter have been generated, which enables time measurements of 0.1 ps or even less to be made. Since the speed of light is

$$c = 299\,792\,458 \text{ m/s} \approx 3 \times 10^8 \text{ m/s} ,$$

the length of a 1 ps laser pulse is only 0.3 mm so that we may say that a thin film of light is flying rather than a light ray progressing.

The peak power of a pulsed laser is very much higher than the output power of a continuous wave (cw) laser. Even for a small laser the peak power is 1 MW or more, and for a large laser it is 1 GW or more. If it is further amplified by laser amplifiers, a peak power of as much as 1 TW is obtainable. Therefore, when such a high-power laser pulse is focused, we can obtain an extraordinarily high energy density both spatially and temporally.

However, the more the light energy is concentrated in time the more it widens spectrally. The frequency spectrum of a light pulse whose envelope is of a Gaussian shape has a Gaussian spectral distribution about the central frequency. If we denote the full width at half maximum of the Gaussian pulse by τ_p and that of the spectrum by $\delta\nu_p$ we obtain the relation

$$\tau_p \delta\nu_p = 0.44 \ .$$

Thus the spectral width of a 1 ps laser pulse is as broad as 440 GHz, corresponding to the frequency of submillimeter waves.

1.2 Solid-State Lasers

A solid-state laser is one in which the laser material is a crystal or glass which has a sharp fluorescent spectral line. Under strong optical excitation it is used as an oscillator or an amplifier at the fluorescence wavelength. Semiconductor lasers and plastic lasers, though solid in material, are not usually included in the category of solid-state lasers [1.3].

The first solid-state laser to be operated successfully, in June 1960, was a ruby laser. Solid-state laser materials are transparent, heat resistant, and hard crystals or glass, containing transition metals or rare earth elements as active ions. Ruby crystals contain 0.01–0.1%[1] of Cr^{3+} in Al_2O_3 (sapphire) and are used for generating and amplifying laser light of 694 nm wavelength.

The most practical solid-state laser which is used at present for material processing and machining is the YAG laser which is made of a $Y_3Al_5O_{12}$ (yttrium aluminium garnet) crystal containing 0.1–1% of Nd^{3+} and emitting at a wavelength of 1.06 μm or sometimes at 1.32 μm. There are also crystals, such as $CaWO_4$ and CaF_2, containing trivalent ions like Nd^{3+}, Pr^{3+}, Ho^{3+}, Er^{3+}, Tm^{3+}, and U^{3+}, of which there are about 30 different kinds used as laser materials. In addition, there are a few laser crystals, such as CaF_2 and SrF_2, containing divalent ions like Sm^{2+} and Dy^{2+}. The wavelengths of these lasers are in the near infrared region, 1.0–2.6 μm, but recently, a laser mate-

1 This is the usual expression when the crystal contains 0.01–0.1% of Cr_2O_3 by weight.

rial called YLF ($YLiF_4$, yttrium lithium fluoride) has been developed to be used for some wavelengths from ultraviolet to infrared. Doping with Ce^{3+} gives the 325 and 309 nm lines, Tm^{3+} gives the 453 nm line, and Tb^{3+} gives the 545 nm line. A few other lines are generated by YLF crystals doped with Pr^{3+}, Ho^{3+}, Er^{3+}, and Nd^{3+}, of which the longest wavelength, 3.9 μm, is given by Ho^{3+} [1.4].

A typical arrangement for a solid-state laser is shown in Fig. 1.1. A laser material, such as those described above, is shaped into a cylindrical rod whose ends are ground and polished to be plane parallel. It is then called a laser rod. When the rod is placed between two mirrors facing each other and is strongly irradiated by an intense flash of light from one or more flash lamps surrounding it, a pulse of laser light is emitted. The size of a typical laser rod is 6 cm in length and 6 mm in diameter, but, occasionally, rods shorter than 1 cm or longer than 30 cm are used.

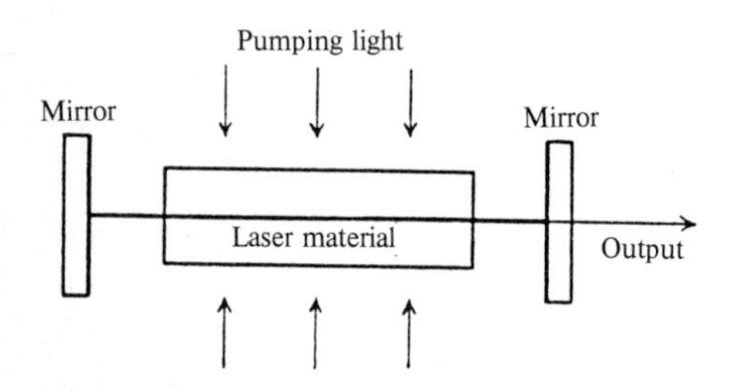

Fig. 1.1. Optically pumped solid-state laser

In a few types of lasers, such as YAGs, continuous waves can be generated by excitation from a continuously radiating light source. An ordinary YAG laser can generate a cw power of 10–100 W. For a pulse width of 10 ns–1 ms it is possible to generate an output energy of 50 mJ–10 J per pulse at a repetition frequency of 100–1 Hz. Continuous operation with a ruby laser is difficult and the cw output is only about 1 mW, but a peak power of 1–100 MW can be obtained depending on the pulse width (10 ns–1 ms) and the quality of the ruby rod.

For a glass laser, silicate glass, phosphate glass, or fluoride glass, containing 1–5% Nd^{3+}, is used; the wavelength of the laser is 1.05–1.06 μm. It is extremely difficult to grow big crystals of ruby or YAG but with glass it is comparatively easy to make a large piece of laser material of optical homogeneity. For this reason glass lasers are used for laser amplifiers of high output power such as are used in the energy driver of inertial confinement nuclear fusion. However, because of the low thermal conductivity of glass, the pulse-repetition rate must be lowered considerably.

Recently, research has been carried out on a small-size solid-state laser, with Nd^{3+} present not as an impurity of low concentration but as a main constituent, for high pulse-repetition rate or cw operation. Crystals of NPP (neodymium pentaphosphate, NdP_5O_{14}) and LNP (lithium neodymium phosphate, $LiNdP_4O_{12}$) are made into platelets thinner than 1 mm and excited with an argon laser to produce a laser output of about 1.05 μm wavelength.

Although the wavelength can be changed within only 1% for the majority of lasers, solid-state lasers which are tunable over a much broader range have recently been developed. The alexandrite laser is made of a $BeAl_2O_4$ crystal with Cr^{3+} impurity and has an output power and an efficiency comparable with those of the YAG laser while possessing a tuning range of 0.70–0.82 μm. It has been found that the output power increases with increasing temperature, up to a temperature somewhat above 200 °C. In the near infrared range, MgF_2 crystals with Ni^{2+} or Co^{2+} impurities have been developed as lasers tunable in the range 1.6–2.0 μm and 1.6–2.3 μm, respectively. In addition, by using several kinds of color-center lasers which are made of alkali-halide crystals, it is possible to cover a wavelength range of 0.9–3.3 μm. Commercial color-center lasers are used in spectroscopic research.

1.3 Gas Lasers

The first laser that generated cw output, in December 1960, was the 1.15 μm He-Ne laser, excited by an rf discharge through a mixture of helium and neon gas. Since then, a large variety of gas lasers has been developed in a wide range of wavelengths, from 100 nm in the vacuum ultraviolet to the far-infrared or even millimeter waves; a total of over 5000 laser lines are known [1.5]. In most cases the gas is either excited by an electric discharge or optically pumped, and a great number of methods and excitation mechanisms are involved.

Gas lasers are classified by their medium as follows: neutral atomic lasers, ion lasers, molecular lasers, excimer lasers, etc. Classifying them by the excitation method, there are electric discharge lasers, chemical lasers, optically pumped lasers, electron-beam-pumped lasers, etc. Excitation by electric discharge may further be classified into direct-current discharge, high-frequency discharge, pulsed discharge and hollow-cathode discharge. With regard to excitation mechanisms, many processes are known, such as electron collision, energy transfer by collision with excited atoms, dissociation of molecules, recombination of ions and electrons, resonance absorption, trapping of resonance radiation, etc.

1.3.1 Atomic Gas Lasers

The most typical of the neutral atomic gas lasers is the He-Ne laser, whose energy levels are briefly shown in Fig. 1.2. When a He atom, excited to the metastable state 2^1S or 2^3S by electric discharge, collides with a Ne atom in

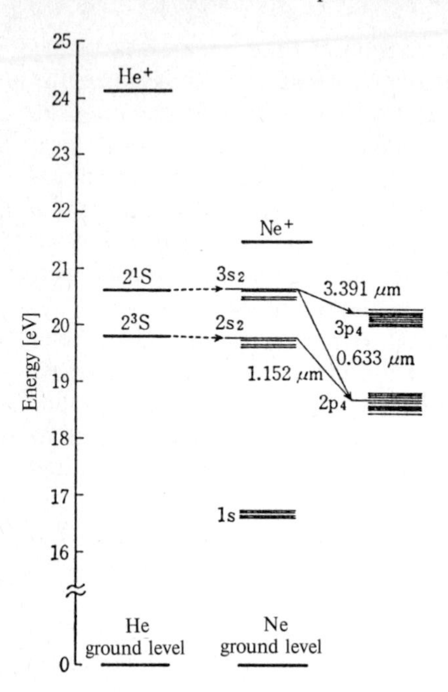

Fig. 1.2. Energy-level diagram of the He-Ne laser[2]

the ground state, the excitation energy of He is transferred to Ne, as a result of which the distribution of Ne atoms in the 2s and 3s levels increases. The transitions from these levels to lower levels result in the laser action of 1.15, 0.63, 3.39 μm and a few other lines in their neighborhood. Typical characteristics of these lasers are shown in Table 1.1. Although the power output of the short 633 nm He-Ne laser is small, it is the cheapest and handiest of the visible (red) lasers; thus, it is widely sold on the market for optical experiments and measurements as well as for display purposes.

Table 1.1. He-Ne lasers

Wavelength [μm]	Transition (Paschen notation)	Length [cm]	Diameter [mm]	Output power [mW]
0.6328	$3s_2 \rightarrow 2p_4$	15	1	1
0.6328	$3s_2 \rightarrow 2p_4$	180	3	80
1.1523	$2s_2 \rightarrow 2p_4$	100	7	20
3.3913	$3s_2 \rightarrow 3p_4$	100	3	10

2 ^{1}S denotes a singlet state for total electron spin momentum $S = 0$ and orbital angular momentum $L = 0$, while ^{3}S denotes a triplet state for $S = 1$ and $L = 0$. States for $L = 1, 2, 3 \ldots$ are denoted by P, D, F... respectively [1.6].

Pure Ne is known to have many laser lines up to the far infrared. Many laser lines have also been found in discharge-excited He, Ar, Kr, and Xe gases, but they are not of much practical use except for spectroscopic research.

What is commonly known as the argon laser does not work with neutral argon atoms, but is an argon ion laser. An intense discharge of several tens of amperes through a laser tube enclosing argon gas delivers more than 20 laser lines of the argon ion spectrum[3], such as the 514.5 and 488.0 nm lines, from green to the ultraviolet. The output power of the stronger lines is over 10 W, and even for the weaker lines it is about 100 mW. The similar Kr ion laser delivers about 20 laser lines ranging from the red 647.1 nm line to the violet. Many of the ion lasers on the market have an output power of 1–10 W. They are used for excitation of dye lasers, to be described later, and small solid-state lasers, described above, as light sources for various spectrometers, and for medical applications.

Laser action can also be obtained from metallic atoms, such as mercury , and their ions by heating them to a high temperature or by the use of a hollow-cathode discharge. These are called metal vapor lasers, but, in spite of their high efficiency in the visible and near ultraviolet range, most of them are still at the development stage because of the difficulty in maintaining the metallic vapor and the discharge for a long time. The He-Cd laser, generating cw radiation at 441.6 and 325 nm from Cd^+, is already in practical use to some extent. Also known are lasers of Hg^+, Se^+, Zn^+, Sn^+, Cu^+, Ag^+, Au^+, In^+, Pb^+, Ge^+, Al^+, Tb^+, etc. Among the neutral metallic atom lasers, a peak output power of 1 kW or more can be obtained from lasers of Pb, Mn, Cu, etc. Even for Pb, however, heating to over 800 °C is necessary, while Mn and Cu must be heated to over 1400 °C. With a Cu laser, cw output of 10–100 W with 1% efficiency is obtainable, and, on account of its yellow (578.2 nm) and green (510.6 nm) colors, it has wide applications. There is a Cu vapor laser that can be operated below 500 °C by dissociating certain chlorides in a discharge, but its output power and life are somewhat less.

Laser action can be obtained with most of the non-metallic elements. When an appropriate gas molecule is dissociated, it produces atoms in the excited state from which laser emission may be obtained. Thus, lasers of O, Br, I, Cl, F, C, N, S, As, etc., can be realized. Among them, the 1.31 μm laser of I atoms produced by dissociating CH_3I or C_3F_7I has been developed in the U.S.S.R. and in West Germany to have a peak power of over 100 GW, for nuclear fusion research [1.7].

Of the 91 elements which exist in nature, the only ones for which laser action have not so far been obtained in their gaseous atoms are, in addition to Li and Ga, the elements Fe, Ni, Ti, and W, etc., whose boiling points are

---

3 Neutral Ar, Ar^+, Ar^{2+} are, respectively, expressed in spectroscopical notation as ArI, ArII, ArIII.

above 2500 °C, and the radioactive elements Po, At, Rn, etc., for which only very small quantities of pure gas can be obtained. Thus, we may conclude that laser action is not limited to special elements, but, in principle, may take place with all kinds of elements.

1.3.2 Molecular Lasers

Since a polyatomic molecule, unlike an atom (monatomic molecule), possesses many vibrational and rotational levels, a molecular spectrum consists of an almost countless number of lines. The kinds of gaseous molecules which serve as laser media are many, but the majority of them are relatively light, small molecules. Of the different kinds of molecular lasers, we shall describe below some of the typical ones.

The N_2 laser is known as an ultraviolet laser, emitting the 337 nm line and excited with short, pulsed discharges. In addition, it has about 400 lines from 0.31 μm in the ultraviolet to 8.2 μm in the infrared. Normally, the pulse width is a few ns and the peak power for the 337 nm line is about 100 kW in a small-size laser and over 10 MW for a large-size one. Comparable, but somewhat less, power may be obtained at 316 nm and at 358 nm. A high-voltage power supply of 30–40 kV is necessary to excite the N_2 laser. Its cw operation is not possible because of the properties of the energy levels.

The H_2 laser, like the N_2 laser, is excited by a pulsed discharge to emit in the far ultraviolet between 110 and 162 nm. Since 1970, this has been the laser with the shortest wavelength, but it has little practical use because the peak power is only 10–100 W.

The carbon dioxide laser gives a cw output at 10 μm in the infrared with a high efficiency and it is the most practical molecular laser. There are a large number of CO_2 lasers, varying in structure, method of excitation, and capacity. Including those of the isotopic molecules ($^{13}CO_2$, etc.) there is a total of nearly 1000 laser lines, the main ones being between 9 and 11 μm. The usual way of obtaining single-line oscillation is to use a diffraction grating in conjunction with a laser resonator. If only mirrors are used, simultaneous oscillation on several lines in the neighborhood of 10.6 μm is commonly obtained. The output power of even a small CO_2 laser is about 1 W and large ones give over 10 kW; these are used for many applications, such as laser material processing and laser heating.

Since CO_2 is a triatomic linear molecule, it has three vibrational modes, ν_1, ν_2, ν_3, as shown in Fig. 1.3: ν_1 is a symmetrical stretching vibration (expansion and contraction), ν_2 is a doubly degenerate bending vibration, and ν_3 is an antisymmetrical stretching vibration. We shall denote the quantum numbers of the respective vibrational modes as v_1, v_2, v_3 and the vibrational state as $(v_1\ v_2\ v_3)$. The ground vibrational state (0 0 0) and the relevant energy levels in the CO_2 laser are shown in Fig. 1.4. In this figure the values of energies and transition frequencies are given in wavenumbers (cm^{-1}), a

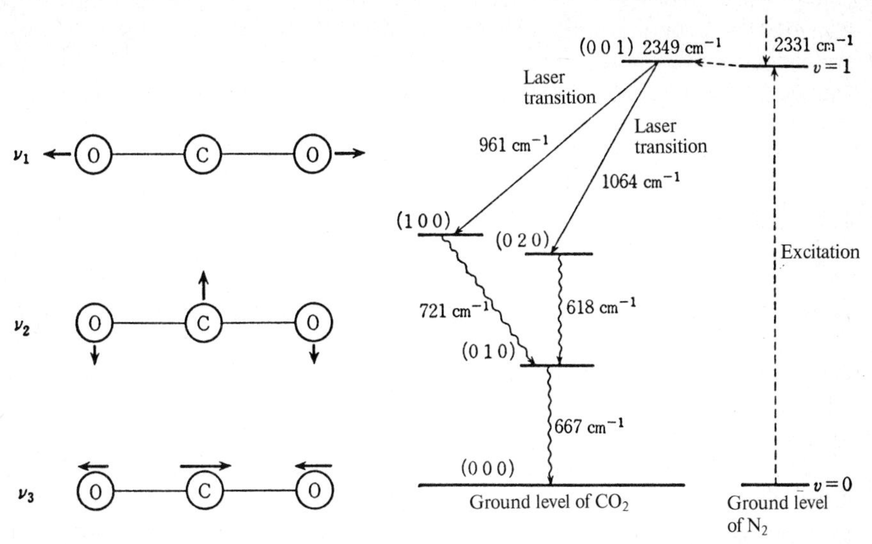

Fig. 1.3. Three modes of vibration in CO_2

Fig. 1.4. Energy-level diagram of CO_2

unit commonly used in spectroscopy. The frequency is about $\tilde{\nu} \times 30$ GHz and the energy is $\tilde{\nu} \times 1.9864 \times 10^{-23}$ J, where $\tilde{\nu}$ is the wavenumber.

Although laser radiation is obtainable with pure CO_2, the usual CO_2 laser uses a mixture of He, N_2, and CO_2. The excited vibrational state of N_2 is metastable and its energy is very close to that of the ν_3 vibrational state of CO_2; thus, resonant energy transfer takes place and an enhanced excitation is obtained. Since the vibrational energy levels of a molecule are almost equally spaced, the higher-excited vibrational levels of N_2 and CO_2 also contribute effectively to the laser. It is believed that the role of He in the medium is to stabilize the discharge and to depopulate the lower levels (1 0 0) and (0 2 0) of the laser transitions.

As the figure shows, the laser transitions are the 961 cm^{-1}, (0 0 1) $\rightarrow$ (1 0 0) transition of the 10.4 μm band and the 1064 cm^{-1}, (0 0 1) $\rightarrow$ (0 2 0) transition of the 9.4 μm band. If the rotational quantum number is denoted by J, the P(J) line of the P-branch corresponds to the $J-1 \rightarrow J$ transition, and R(J) of the R-branch corresponds to the $J+1 \rightarrow J$ transition. Owing to the symmetry of the CO_2 molecule, laser transitions occur to lower energy levels whose rotational quantum numbers J are even, resulting in more than 30 laser lines in each branch.

The normal cw CO_2 laser is excited by direct current with a total gas pressure of 10^2–10^3 Pa (1 Pa = 7.5 mTorr). With the pressure raised to 10^4–10^5 Pa, close to atmospheric pressure, a peak power of 10 to over 100 MW is obtainable by using a pulsed discharge through a number of short gaps arranged along the tube axis perpendicular to it. This laser is called a transversely excited atmospheric laser, abbreviated to TEA laser. Some have needle electrodes and others have rod or plate electrodes. The main discharge is controlled by triggering with a preceding discharge or an electron beam.

Although N_2O and CO lasers have a lower output power than the CO_2 laser, they have nearly 100 laser lines each in the ranges 10–11 μm and 5–6.5 μm, respectively (considering the main isotopic species, i.e., $^{14}N_2{}^{16}O$ and $^{12}C^{16}O$, only). CO laser with high output power has recently been developed, whose efficiency, like that of the CO_2 laser, has reached a value of 10–30%. Besides these polyatomic molecules, the molecules NH_3, OCS, CS_2, etc., also have quite a few laser lines in the infrared. Furthermore, with SO_2, HCN, DCN, H_2O, D_2O, etc., many laser lines are obtained in the far-infrared from 30 μm up to submillimeter wavelengths.

Optical excitation of molecules such as CH_3F, CH_3OH, and HCOOH, with the output of a CO_2 laser at a resonance-absorption wavelength of these molecules, gives rise to laser emission in the far-infrared. This type of laser is known as an optically pumped far-infrared laser. At present, over 1000 such laser lines are known, most of which are obtained only in pulses with moderate output powers but some of which deliver a high peak output power with a fairly high efficiency. Quite a number of them may also operate cw if excited continuously. These are employed for research in the far-infrared where appropriate light sources have not been available before.

The excimer laser has recently been developed as a laser of large output power with high efficiency in the short-wavelength range [1.8]. Although an excimer in the ground state is unstable, it is a stable diatomic molecule in an electronic excited state. The energy levels of the Xe_2 laser, which first gave emission in 1970 at a wavelength of 173 nm, are given in Fig. 1.5 as a function of atomic separation.

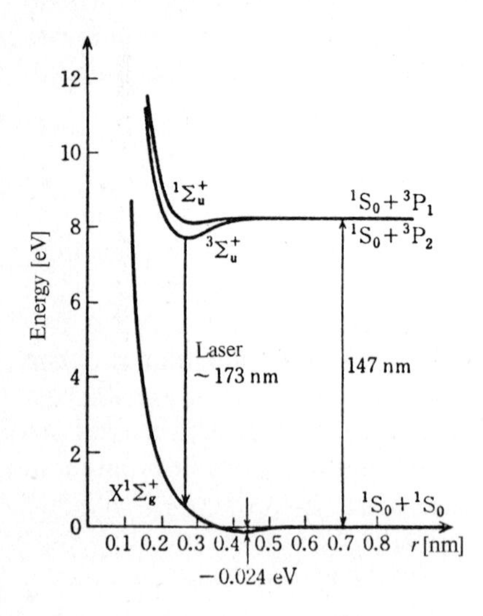

Fig. 1.5. Energy level diagram of Xe_2. X denotes the electronic ground state. Σ denotes the electronic state where the component of the total electronic orbital angular momentum of a diatomic molecule along the direction of the molecular axis is zero. The suffix g (gerade) means that the orbital wave function is unchanged on inversion of coordinates with respect to the center of the molecule; the suffix u (ungerade) means that the sign changes with inversion; the right superscript ± denotes the symmetry for mirror reflection with respect to a plane containing the molecular axis

Other rare gas excimer lasers have similar energy levels. The excimer laser may be excited either by an electron beam or by an electric discharge. A peak power of 10–100 MW in the vacuum ultraviolet is obtained at 146 nm with Kr_2 and at 126 nm with Ar_2 under pulsed excitation.

Subsequently, excimer lasers of rare gas halides have been developed with still better properties. Excimers such as XeCl, KrF, ArF, formed of different atoms, are called heteroexcimers or exciplexes. These heteroexcimer lasers have been studied and developed since 1975 as pulsed lasers at 308, 249, and 193 nm, respectively. They emit pulses of a few ns in duration with a peak power of at least 10 MW, and more than 1 GW for a scaled-up system. Excimer lasers are being actively used in photochemical research. Moreover, research is also in progress on their use for inertial confinement nuclear fusion.

A typical example of chemical lasers is the HF laser with a wavelength of about 2.7 μm and a high output power. It utilizes a chemical reaction between H_2 and F_2, forming HF molecules in excited vibrational states to give rise to laser action. Actually, in order to control the reaction, F_2 is replaced by SF_6, or a small quantity of O_2 is added. The reaction is initiated by a pulsed discharge or by electron-beam irradiation. When an F atom is liberated, the exothermal reactions

$$F + H_2 \rightarrow HF^* + H$$

$$H + F_2 \rightarrow HF^* + F$$

progress with the formation of many HF molecules in excited vibrational states (* denotes an excited state), but with scarcely any formation of molecules in the ground state. Consequently laser radiation is emitted mostly on the $v = 3 \rightarrow 2$, $2 \rightarrow 1$, and $1 \rightarrow 0$ transitions. Since the chemical reaction propagates at a speed of the order of that of sound waves, the laser pulse is of the order of μs in duration and the output energy ranges from 1 J to more than 1 kJ per pulse. The DF and HBr chemical lasers emit longer wavelengths than the HF laser, and their output power is lower.

1.4 Dye Lasers

Laser action can be attained in an optically pumped dilute solution of an organic dye in ethanol, cyclohexane, toluene, etc. This is called a dye laser [1.9], and, since the fluorescence spectrum of a dye is as wide as 100–200 cm^{-1}, the laser wavelength can be tuned with ease by the use of a diffraction grating or a prism. The light source for optical pumping is either a flash lamp, an Ar or Kr ion laser, an N_2 laser, a KrF laser, a Cu laser or higher harmonics of a solid-state laser. Since the success of a pulsed dye laser in 1966, lasers have been made with over 500 different dyes [1.10]. But, relatively few of these dyes allow cw operation. Practically, about 10 different dyes are enough to obtain laser output of any wavelength in the visible range. An example is given in Fig. 1.6 for dyes optically pumped by an N_2 laser.

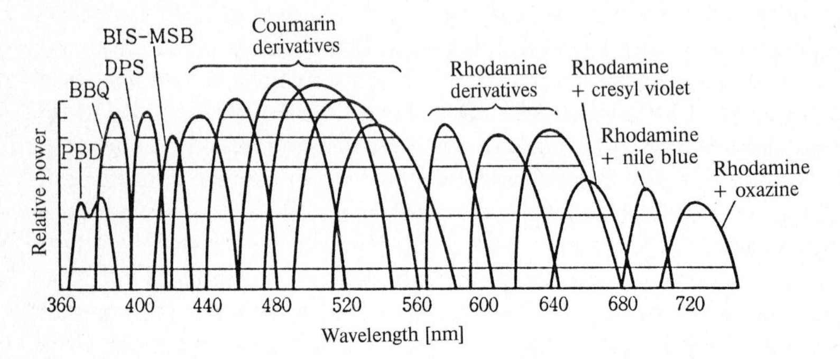

Fig. 1.6. Wavelength and output power of a series of dye lasers excited by an N_2 laser (Molectron Corp.)

The efficiency of a dye laser is relatively high but the dye gets heated by the pumping light and laser action degrades; in order to achieve cw operation, therefore, the dye solution must be cooled by circulation. Optical loss at the windows of the dye cell is usually eliminated by using a ribbon-shaped jet flow. The cw output thus obtained may exceed 1 W, but usually it is 10–100 mW. The peak power of a pulse is higher than 1 kW and can be amplified to above 100 MW. Also, by exciting a dye laser with a train of pulses from a mode-locked laser (Sect. 9.4), it is possible to generate synchronized pulses of an extremely short duration, the shortest being less than 0.1 ps long [1.2].

Since the dye laser medium is a liquid, it is inconvenient to handle, and its use is somewhat limited. However, a variety of commercial dye lasers are on the market for use in spectroscopic research and spectrum analysis, since they are the most convenient tunable lasers in the visible.

1.5 Semiconductor Lasers

Although a semiconductor laser can be operated with optical or electron-beam excitation, its distinctive features are that it lases with a current through a *p-n* junction and is extremely small in size compared to other lasers. A semiconductor laser had been proposed in 1957 but it was not till 1962 that the laser was first demonstrated, in pulsed operation at low temperature. Since cw operation was successfully achieved at room temperature in 1970, it has made rapid progress as the light source for optical fiber communication [1.11].

The energy levels of a crystal are in the conduction band and the valence band, respectively above and below the forbidden band. In a semiconductor, however, there are few electrons in the conduction band while the valence band is almost filled with electrons. This makes electric conduction difficult, since the carriers of electric current, namely, electrons and positive holes, are few. In *n*-type semiconductors containing some impurities, however, a considerable number of electrons are excited from the impurity levels to the conduction band, while in *p*-type semiconductors containing other impurities electrons in the valence band are excited to the impurity levels, resulting in a considerable number of positive holes: in either case comparatively good electric conduction is realized.

The so-called *p-n* junction is made of p-type and *n*-type regions in a single crystal of a semiconductor [1.12]. Very little current flows when the *p*-type is negatively biased with respect to the *n*-type; but if the *p*-type is positively biased, current will flow easily. In the latter case, the positive holes in the *p*-type region are injected into the *n*-type region, while electrons in the *n*-type are injected into the *p*-type region. When an electron meets a positive hole, they combine together, emitting a photon of energy nearly equal to the band-gap energy. The light-emitting diode (LED) is a tiny light source based on this phenomenon. The recombination of an electron and a positive hole with the emission of a photon and without the emission of a phonon is called a direct transition, and the simultaneous emission of a photon and a phonon, in order to conserve momentum, is called an indirect transition.

The structure of a simple semiconductor laser is shown in Fig. 1.7. The two parallel facets[4] of a direct-transition semiconductor are perpendicular to the plane of the *p-n* junction, where a positive potential is applied to the electrode on the *p*-type semiconductor and a negative to the *n*-type. For example, if a current of either a few tens of kA cm^{-2} at room temperature or a little over 1 kA cm^{-2} at liquid nitrogen temperature is run through the *p-n* junction of GaAs, we obtain laser oscillation at wavelengths of about 0.85 μm. The directivity of the output from a semiconductor laser is not sharp, spreading out by 5°–30°; this is because the light is emitted from a small area whose sides are about 2–20 μm. The intensity of emitted light as a function of injection current is shown in Fig. 1.8. It is seen that the intensity increases rapidly above a certain current I_{th}, while below I_{th} it is rather weak. This current I_{th} is the starting current for laser oscillation and is called the threshold current. When the current is below threshold, the directivity of the emitted light is poor and the spectral width is broad, being rather similar to a light-emitting diode. Above threshold, on the other hand, the laser light has a narrower spectral width and sharper directivity.

4 The cleaved face is normally used.

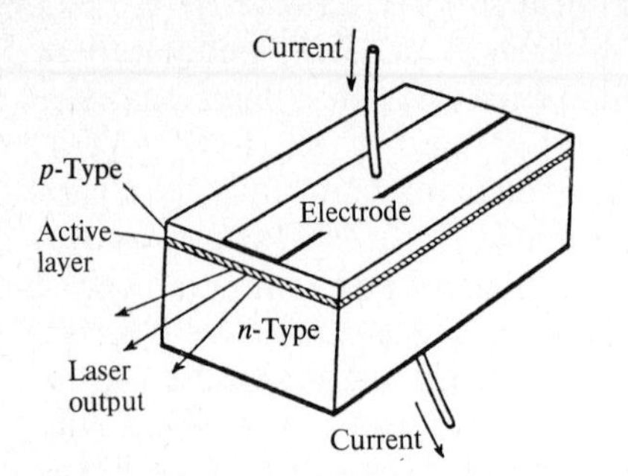

Fig. 1.7. Simple semiconductor laser

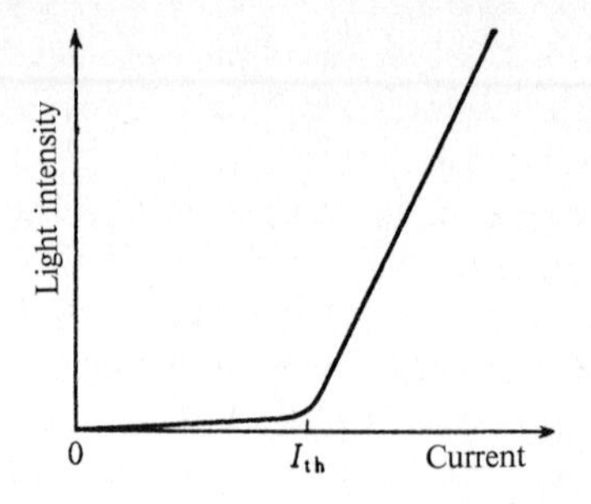

Fig. 1.8. Light output vs. excitation current of a semiconductor laser

We cannot obtain a laser from Si or Ge because the direct transition does not occur in them, but with InP, InAs, and InSb we have lasers at wavelengths of 0.91, 3.1, and 5.2 μm, respectively. The lasing wavelength varies slightly with temperature and current. In particular, the wavelength may change by as much as a few tens of percent with external magnetic field and pressure in some cases. If a *p-n* junction is made from a mixed crystal, such as $In_xGa_{1-x}As$, made up of two different materials, InAs and GaAs, it will lase at a wavelength tunable between 0.9 and 3.1 μm depending on the mixing ratio *x*. The tuning of various tertiary semiconductors is given in Fig. 1.9.

The *p-n* junctions of the early semiconductor lasers were made from a single material and were known as homojunctions. But the recent ones are

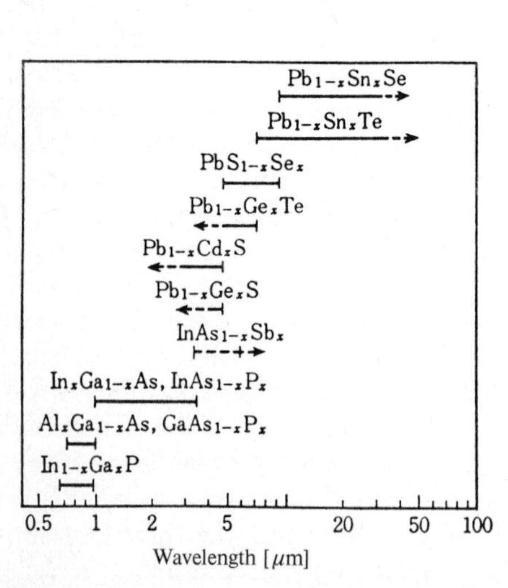

Fig. 1.9. Wavelength range of output of various tertiary semiconductor lasers

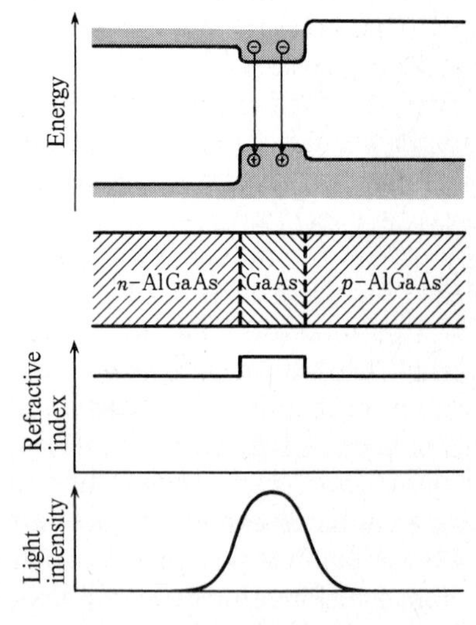

Fig. 1.10. Energy level, refractive index and light intensity distribution of a DH-structure laser

made from junctions of two different kinds of semiconductors and are called heterojunctions. Here, on comparing GaAs with $Al_xGa_{1-x}As$ (hereafter abbreviated to AlGaAs), we find that the band-gap energy of AlGaAs is greater than that of GaAs, and the refractive index is smaller. Therefore, in a heterojunction of GaAs and AlGaAs, the light is concentrated in GaAs, whose refractive index is greater (Sect. 3.6), and, moreover, carrier electrons and positive holes are also concentrated in GaAs because the band gap is narrower. Thus, if we make a heterojunction of *p*-type AlGaAs and *n*-type AlGaAs on either side of a thin active layer of GaAs, as shown in Fig. 1.10, both light and carriers are confined around the active layer. Not only does such a device lead to a strong interaction between the light and the carriers but it also reduces the unnecessary losses of both. Thus the threshold of laser oscillation is appreciably reduced. Such a *p-n* junction is called a double heterojunction or double heterostructure (abbreviated to DH structure). Since the threshold current density for a GaAs-AlGaAs laser of DH structure is below 1 kA cm^{-2} even at room temperature, cw operation can be obtained with appropriate cooling.

Now, it is necessary that the lattice constants of the two different semiconductor crystals in a DH structure should fit as closely as possible. Otherwise, the lifetime of the carriers becomes short owing to the lattice defect, and not only does this raise the threshold current, it also leads to a rapid degradation of the laser. For a DH-structure laser of 1.2–1.7 μm wavelength it is not possible to obtain a lattice-constant fit with a tertiary compound. A quarternary compound $Ga_xIn_{1-x}As_yP_{1-y}$ is therefore used. The wavelength can be varied by changing either x or y, but, in order to match the lattice constant with the substrate (usually InP), both x and y are fixed. In a DH-structure laser with a planar structure, as shown in Fig. 1.7, the light in the direction perpendicular to the junction plane is concentrated near the active layer, but the light in the transverse direction is widely spread and its intensity distribution is variable for even a small change in current. In Fig. 1.7, the width of the upper electrode is made narrow into what is called a stripe structure in order to concentrate the laser action in the center. But, this is insufficient to prevent the lasing region from spreading horizontally. Consequently, many different structures for semiconductor lasers have been devised to confine the active area not only to a small depth, but also to a small width in the transverse direction. As an example of such a structure, the cross-section of a buried heterostructure is shown in Fig. 1.11. After making a DH structure of AlGaAs with an active layer of GaAs, only the central part is left, covered by a mask a few μm wide, the rest being removed by etching. The etched parts are replaced by growing a crystal of *n*-type AlGaAs. With this procedure the active layer is surrounded on all sides (above and below, left and right) by AlGaAs, and, since the injection of carriers is also confined to the central part, the mode of laser radiation in the heterostructure is similar to that of microwaves in a waveguide of rectangular cross-section.

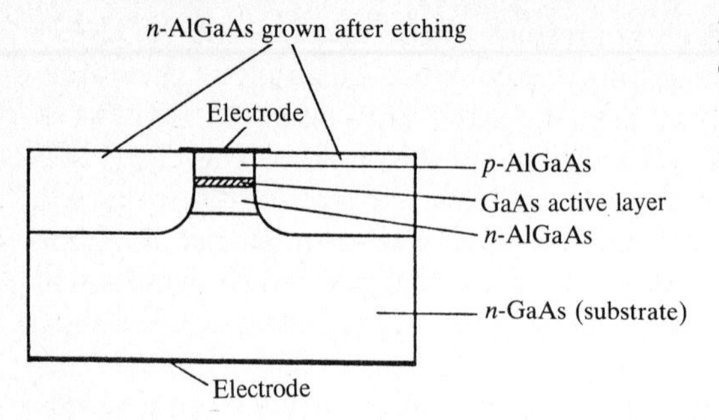

Fig. 1.11. Cross-section of a buried double-heterostructure laser

The length of a semiconductor laser is normally 100–500 μm, the stripe is 2–20 μm wide, the thickness of the active layer is 0.1–2 μm, and a cw power of 1–100 mW is obtained. However, at a wavelength longer than 3 μm the power is smaller. The peak power in pulsed operation is 1–10 W or more, and a pulse width of the order of a pico-second can be obtained.

1.6 Other Lasers and Laser Physics

Quite different methods of excitation are used in some lasers, although their materials are of the same kind as those described above. Without going into detail we list a few of them: excitation by the flash of a wire which explodes when it passes a large current, gas-dynamic excitation by a shock wave or a jet flow, plasma excitation, etc.

Of the lasers with different structures, the important ones are the ring laser, the waveguide-type gas laser, and semiconductor or dye lasers using Bragg reflection or distributed feedback. Many varieties of resonators for solid-state and gas lasers are now being used, such as the composite resonator for improved frequency response and mode characteristics, the unstable resonator which allows remarkably different output characteristics, etc.

Finally, we must mention lasers which differ in principle from those we have described so far, such as the Raman laser and the free electron laser. In the Raman laser use is made of the stimulated Raman effect which takes place when strong laser light is incident on a gaseous, liquid, or solid material: the frequency of the Raman laser is shifted from that of the incident light by the resonance frequency of the material.

The optical parametric oscillator is akin to the Raman laser. If laser light of frequency ν_{p} is incident on a nonlinear optical crystal, the frequency ν_{p} is broken up into the idler frequency ν_{i} and the signal frequency ν_{s} such that $\nu_{\mathrm{p}} = \nu_{\mathrm{i}} + \nu_{\mathrm{s}}$; by the use of an appropriate resonator, it works as an oscillator or amplifier of frequency ν_{s}.

In the free electron laser, a high-speed electron beam is modulated by passing it through a spatially modulated magnetostatic field or a high-frequency electromagnetic field, and it thereby emits coherent light. Since it has the remarkable features that the frequency can be tuned over a wide range and the efficiency is expected to be high with a large output power, rapid progress is being made in its research, both theoretically and experimentally. Because the optical parametric oscillator and the free electron laser do not make use of quantum resonance, there is a point of view that they may not be lasers in the proper sense. However, since the emitted light is coherent and has a common character with laser light, they may surely be regarded as a kind of laser.

Physics of coherent light and laser characteristics are described in the following chapters. In addition, there are many new phenomena and practical problems in laser physics. We should remind ourselves that the interaction between strong coherent light and atoms is the fundamental process of laser physics. Based on the treatments in this book, therefore, it should be possible to understand such problems as the dynamics of mode control, Zeeman polarization, non-steady oscillation and other uncommon characteristics of the laser, as well as nonlinear optical effects and their practical use.

In Sect. 9.6 we give an outline of the fully quantum-mechanical theory of the laser which, in some sense, is the most fundamental theory of laser physics. Although we leave out such typical nonlinear optical effects as optical harmonic generation, stimulated scattering, and self-focusing of light, they should be understood from straightforward extension of the semiclassical treatments described in Chaps. 7–9. Remarkable problems of laser physics in recent years include various types of instability in laser oscillation, optical bistability and its applications, superradiance, multi-photon processes, and many other higher-order optical effects in matter.

Problems

1.1 What kinds of solid-state lasers are mentioned in this chapter which operate in the visible range?

Answer: Ruby, alexandrite, YLF: Tm^{3+}, YLF: Tb^{3+} lasers.

1.2 Find as many lasers as possible that can generate more than 1 W of cw power in the visible range.

Answer: Alexandrite, Cu vapor, Kr^{+}, Ar^{+}, dye lasers.

1.3 Which laser has the highest peak power, respectively, in the following wavelength ranges?

a) 100 to 1000 μm
b) 10 to 100 μm
c) 2 to 10 μm
d) 0.5 to 2 μm
e) 0.1 to 0.5 μm

Answer: It is not quite definite but (a) CH_3F, (b) CO_2, (c) HF, (d) Nd-glass, (e) KrF lasers, respectively.

1.4 Calculate the average power density of a 10 mW semiconductor laser in a 0.1 μm × 2.0 μm cross-section. Compare the value with that of a pulsed YAG laser of 6 mm diameter with a peak power of 1 MW.

Answer: 5×10^{10} W m^{-2}, 3.5×10^{10} W m^{-2}.

1.5 What kinds of semiconcuctor can be used in a laser diode operating at a visible wavelength?

Hint: See Fig. 1.9.

1.6 Calculate the maximum storable energy density for the 694 nm emission in ruby with a Cr^{3+} concentration of 0.02 wt.%. Ruby has a density of 4 g cm^{-3}.

Answer: 6.3×10^{18} photons cm^{-3}, 1.8 J cm^{-3}.

2. The Coherence of Light

Interference and diffraction of light are phenomena known from various experiments. Whereas a dark room used to be necessary to observe interference and diffraction with the use of a monochromatic spectral light source, it is now possible to observe them in an ordinary room in daylight by using a laser light source. The phenomena of interference and diffraction of light are due to the wave nature of light, which will be discussed in this chapter.

2.1 Young's Experiment

Perhaps the most typical of the experiments showing the wave nature of light is that which was first performed by *T. Young* in 1801 and is known as Young's experiment.

As shown in Fig. 2.1, a narrow slit P_0 is placed near a light source S. When the light diffracted through this slit passes through the two slits P_1 and P_2 in the baffle B, fine interference fringes are formed on the screen C at the right of the figure. The spacing of these fringes is explained as follows.

As shown in Fig. 2.2, the x axis is taken in the direction from the slit P_0 to the screen. The screen and the slits are perpendicular to the x axis. We take the first slit P_0 to be on the z axis. The two slits P_1 and P_2 are parallel to the z axis at $x = l_0$, $y = \pm a$. We now consider the interference of light emerging from P_0, passing through P_1 and P_2, and arriving at the point Q $(l_0 + l, y)$ on

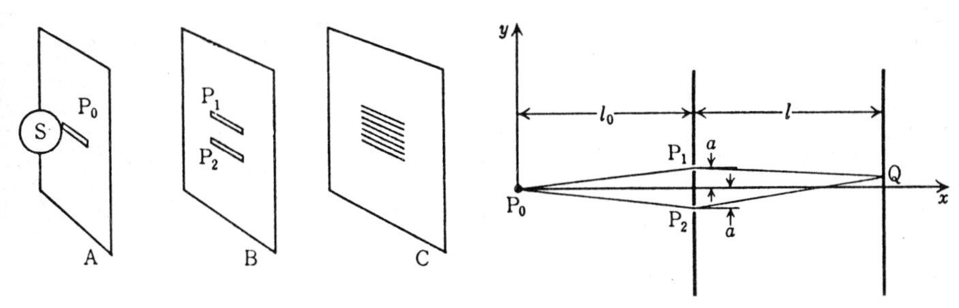

Fig. 2.1. Young's experiment. S is the light source, P_0, P_1, and P_2 are slits, and C is the screen

Fig. 2.2. Two optical paths in Young's experiment

the screen placed at $x = l_0 + l$. The optical path length from P_0 to Q, passing through P_1, is

$$s_1 = \overline{P_0P_1} + \overline{P_1Q} = \sqrt{l_0^2 + a^2} + \sqrt{l^2 + (y-a)^2}\ ,$$

while the optical path length from P_0 to Q, passing through P_2, is

$$s_2 = \overline{P_0P_2} + \overline{P_2Q} = \sqrt{l_0^2 + a^2} + \sqrt{l^2 + (y+a)^2}\ .$$

The difference in the two optical path lengths becomes

$$s_2 - s_1 = \sqrt{l^2 + (y+a)^2} - \sqrt{l^2 + (y-a)^2}\ .$$

If the screen is sufficiently far away so that $l \gg a$ and $l \gg y$, this may be approximated as

$$s_2 - s_1 = \frac{2ay}{l}\ . \tag{2.1}$$

If the amplitude of light arriving at Q from P_1 and P_2 is now denoted by A_1 and A_2, respectively, and the circular frequency[1] is denoted by ω, the superposition of the amplitudes of the two light waves can be expressed as

$$A_1 \cos(\omega t - ks_1) + A_2 \cos(\omega t - ks_2)\ ,$$

where the phase at P_0 at time $t = 0$ has been taken equal to zero. The wavenumber[2] k is given by

$$k = \frac{\omega}{c} = \frac{2\pi}{\lambda} \tag{2.2}$$

where c is the velocity of light and λ the wavelength.

Since the light intensity I at the point Q can be expressed as the time average of the square of the amplitude, we see that

$$I = \tfrac{1}{2}A_1^2 + \tfrac{1}{2}A_2^2 + A_1A_2 \cos k(s_2 - s_1)\ , \tag{2.3}$$

since the time average of $\cos^2 \omega t$ is equal to 1/2. It is seen from (2.3) that the light intensity is a maximum when $k(s_2 - s_1) = 2n\pi\,(n = 0, \pm 1, \pm 2, \ldots)$ and a minimum when $k(s_2 - s_1) = (2n + 1)\pi$. In other words, we have brightness when $s_s - s_1 = n\lambda$ and darkness when $s_2 - s_1 = (n + 1/2)\lambda$. If these are substituted in (2.1), the position y_n of the nth bright fringe on the screen is given by

$$y_n = n\frac{l\lambda}{2a}\ , \quad n = 0, \pm 1, \pm 2, \ldots\ .$$

1 We will hereafter write simply frequency instead of circular frequency.

2 In order to discriminate against the spectroscopic wavenumber given by $1/\lambda$, k is called the wavelength constant or the phase constant.

The fringes are equally spaced by an amount

$$y_{n+1} - y_n = \frac{l\lambda}{2a} \,. \tag{2.4}$$

Since the experimentally observed fringe spacing is directly proportional to l and inversely proportional to a, the wavelength of the light can be calculated from (2.4).

In particular, if the widths of the two slits are equal, we have $A_1 = A_2$, and the light intensity on the screen becomes

$$\begin{aligned} I &= A^2[1 + \cos k(s_2 - s_1)] \\ &= A^2\left(1 + \cos\frac{2kay}{l}\right) . \end{aligned} \tag{2.5}$$

Since the light intensity at minimum brightness becomes zero in this case, the interference fringes we observe here have the most distinct difference between brightness and darkness.

2.2 The Michelson Interferometer

In the Michelson interferometer (Fig. 2.3) a parallel beam of light is divided into two beams in such a way that there is an optical path difference when they rejoin, thus resulting in interference. A light beam from a source S is partly reflected and partly transmitted by a half-silvered mirror (or a beam splitter), after which the divided beams are, respectively, reflected by plane mirrors M_1 and M_2. The reflected beams are again partly reflected and partly transmitted by the half-silvered mirror onto a detector D. If the distances from the point where the light beam falls on the half-silvered mirror to the

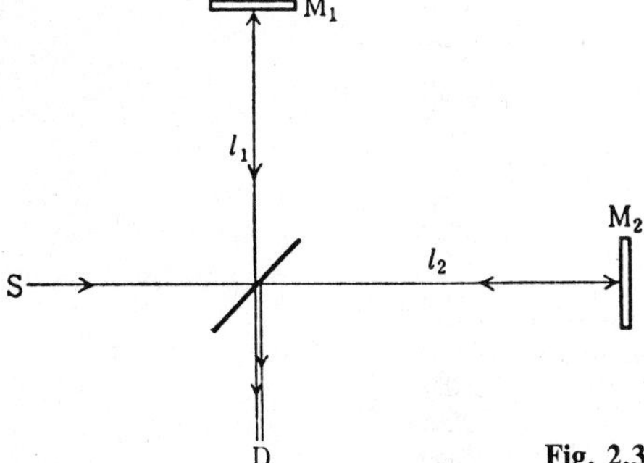

Fig. 2.3. Michelson interferometer

mirrors M_1, M_2 are taken as l_1, l_2, respectively, the optical path difference between the two beams entering the detector D is

$$s_2 - s_1 = 2(l_2 - l_1) \ . \tag{2.6}$$

Now, as we can see immediately from (2.3), the intensity of light entering the detector varies as the mirror M_2 (or M_1) is moved along the direction of the light beam, according to the relation

$$I = \tfrac{1}{2}(A_1^2 + A_2^2) + A_1 A_2 \cos 2k(l_2 - l_1) \ . \tag{2.7}$$

This intensity varies as shown in Fig. 2.4 with a period of half a wavelength. Generally, the light beam entering the interferometer is not perfectly parallel but contains light rays which are slightly inclined to the beam axis. The optical path difference changes gradually as the ray is inclined to the beam axis (Sect. 3.5), so that an interference pattern of concentric circles may be observed at D.

If the amplitudes A_1 and A_2 of the two light beams that rejoin after being divided by the half-silvered mirror are equal, the intensity of the dark ring is zero and we obtain a distinctly clear interference pattern. If, however, we use an ordinary monochromatic spectral light source, the interference pattern observed as the mirror M_2 is moved is very clear while the optical path difference is small, but as the optical path difference is increased to a distance of several centimeters, the interference pattern becomes more and more indistinct and finally disappears completely. The limiting optical path difference over which the interference pattern is tolerably distinct for that particular wavelength is called its coherence length.

In order to express the distinctness of the interference pattern quantitatively, we define the quantity

$$V = \frac{I_{max} - I_{min}}{I_{max} + I_{min}} \tag{2.8}$$

where I_{max} and I_{min} are, respectively, the intensity maximum and minimum of the fringes, as shown in Fig. 2.4. The quantity V is called the visibility of the fringes.

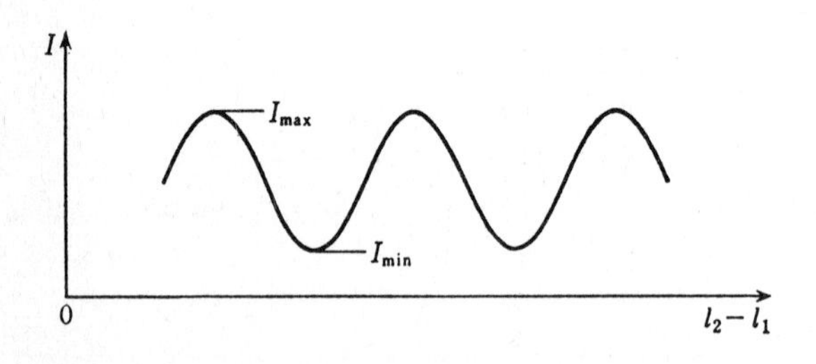

Fig. 2.4. Variation of light intensity entering the detector D as the mirror M_2 is moved

2.3 Temporal Coherence and Spatial Coherence

As the optical path difference $s_2 - s_1$ in the Michelson interferometer is increased, the fringe visibility, in general, decreases as shown in Fig. 2.5. The optical path difference at which the fringe visibility begins to decrease appreciably depends on the quality of the light source, and the coherence length is generally determined without rigorously defining the limiting length of the optical path difference. We shall now consider why the fringe visibility of the interference pattern decreases as the optical path difference is increased.

If a light wave is an infinite continuous plane wave expressed as $A\cos(\omega t - ks)$, where the amplitude A, the frequency ω and the wavenumber $k = \omega/c$ are all assumed constant, the interference pattern should appear as calculated from (2.7), however large the optical path difference may be. The reason why the interference pattern disappears in reality as the optical path difference increases must be attributed to the fact that the light waves emerging from the light source are not a long continuation of harmonic waves, but rather a series of waves of shorter duration. The reason why light from a light source appears steadily bright, on the other hand, is because these short impulsive waves appear one after another and are superposed incessantly.

The energy of light emitted from an excited atom is a constant ($\hbar\omega$), and the corresponding light waves may be considered a damped oscillation, with the amplitude decreasing in time as shown in Fig. 2.6. If τ_a denotes the lifetime of an excited state of an atom, the intensity of the damped light reduces to 1/e in time τ_a, while the amplitude is reduced to $1/\sqrt{e}$; the amplitude will become 1/e of the original amplitude in time $2\tau_a$. Thus, the amplitude $E(t)$ of the damped oscillation beginning at time $t = 0$ can be expressed as

$$E(t) = A\exp(-t/2\tau_a)\cos(\omega t + \theta) , \tag{2.9}$$

where θ is the phase at $t = 0$.

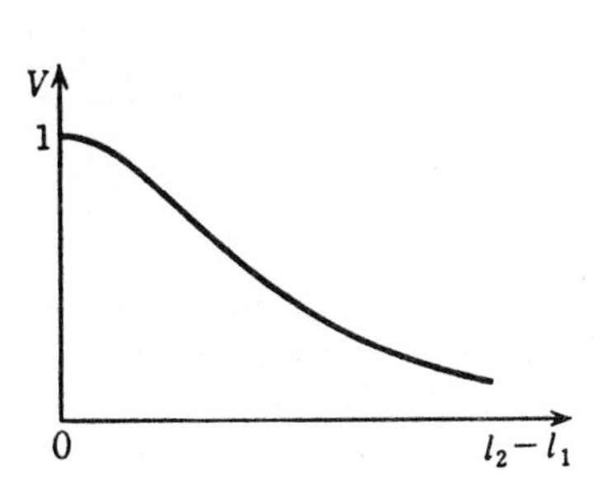

Fig. 2.5. Visibility as a function of the optical path difference (schematic)

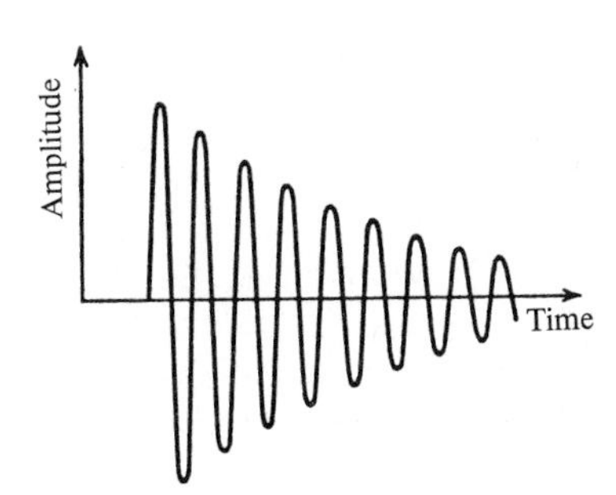

Fig. 2.6. Waveform of a damped oscillation

In an actual light source there are many atoms from which light is emitted. Here, just when and where and with what phase each individual atom emits light is completely random. Consequently, since there are many atoms emitting light at a statistically constant rate, the phases of two light waves which are emitted at times separated by a considerable period are essentially unrelated, even though the light intensity is constant on average [2.1]. This is the reason why we observe a distinct interference pattern in the Michelson interferometer as long as the optical path difference of the two divided beams is less than $c\tau_a$, but we find a disappearance of the interference pattern when the optical path difference is greater; in the latter case we have a superposition of two waves which were emitted at times separated by an interval much greater than τ_a.

When two light beams interfere to give a distinct interference pattern we say that they are coherent, and when they do not give any interference pattern at all they are said to be incoherent. When two light beams interfere to give a fringe visibility of $V = 1$, they are perfectly coherent, whilst if $V = 0$, they are perfectly incoherent. Therefore, the fringe visibility V is used to express how coherent the light is, and is called the degree of coherence.

Laser light has very good interference properties. Rigorously speaking, however, it is not perfectly coherent, so that V is slightly less than 1, although it is almost equal to 1. Even with laser light of rather poor coherence, the coherence length will be at least a few centimeters, and for laser light of good coherence the length will be as much as 1000 km or more. The interference property of an ordinary incandescent lamp is very poor and is almost perfectly incoherent when the optical path difference exceeds a few μm. It is not, however, absolutely incoherent, and, at an optical path difference smaller than a few μm it has a non-vanishing degree of coherence.

According to what we have so far explained, the coherence length of light from a monochromatic spectral light source, that is, an atomic spectral line, should be of the order of $c\tau_a$ or $2c\tau_a$. Here, τ_a is the lifetime of the excited state, and, therefore, is the damping time of emission from each individual atom or the duration of each light pulse. For example, if τ_a is 1 ns $= 10^{-9}$ s, the coherence length should be at least about 30 cm; and if τ_a is 10 ns $= 10^{-8}$ s, it should be at least about 3 m. In reality, however, the coherence length of spectral light is very much shorter than these values. This is not because the lifetime is shorter, but because the optical frequencies emitted from individual atoms in the light source are slightly different from each other. There are various reasons for this; lack of homogeneity of the materials, the effect of magnetic and electric fields, etc., but the main reason, in the case of emission lines of gaseous discharge at low pressures, is the Doppler effect due to the thermal motion of gaseous atoms.

When the emission frequencies of atoms in a light source are not identical but are distributed over a certain frequency range, the interference fringes formed by light from individual atoms will not be uniformly spaced and there will be corresponding displacements among the fringes. Consequently, the fringe visibility of the interference pattern formed by light due to the ensemble of all atoms in the light source will be decreased. When the emission frequency of each atom differs from the ensemble average by $\Delta\omega$, the phase of the light will become practically random in a time t, where $\Delta\omega t > 2\pi$. Even if the damping time τ_a is long, therefore, the coherence will be poor, lasting for only about $2\pi/\Delta\omega$ in time. Thus, although the temporal coherence of light emitted from each individual atom is of the order of the lifetime of the excited state of the atom, the temporal coherence of the spectral line lasts only for a short time of the order of the inverse of the width of the spectral line.

We shall next consider what is known as spatial coherence. Interference fringes appear on the screen in Young's experiment because light coming from two different points in space interferes. When light from two different points in space interferes to give a distinct interference pattern, we say that there exists spatial coherence.

Suppose that the width of the slit P_0 of Young's experiment in Fig. 2.1 is widened; then the fringe visibility of the interference pattern on the screen will be reduced. Why is this? If the width of the slit P_0 is wide, there is practically no diffraction at P_0 and light from the light source reaches the baffle B in a straight line, so that light waves reaching P_1 and P_2 come from two different points of the source. Since different atoms are emitting at different points of the source, they will bring about no interference. Thus, as the width of the slit P_0 is widened the spatial coherence decreases because light waves emitted from different points of such a wide light source are incoherent.

Now, if we use a laser, even though the width of the slit P_0 is widened to the extent of there being virtually no slit at all, we still find interference fringes of high visibility. Not only does laser light have high temporal coherence, it also has very high spatial coherence, so that light from any two separate points on a laser can interfere distinctly. It can be shown, moreover, that two laser beams transmitted in opposite directions from both ends of a laser will exhibit distinct interference when they are superposed on each other by using appropriate reflectors.

The spatial coherence in the direction of propagation is, in principle, determined by the temporal coherence. The spatial coherence length in the direction of propagation (longitudinal direction) for a plane wave of perfectly parallel rays or a spherical wave spreading out by diffraction from a small aperture, is equal to c times the duration of the temporal coherence. If the direction of the ray is not homogeneous and if composite rays of different directions of propagation are involved, we have poorer spatial coherence. It may be said that for ordinary light, there is very little correlation between temporal coherence and spatial coherence in the lateral direction. Consequently, it is generally necessary to distinguish spatial coherence from temporal coherence.

2.4 Complex Representation of Light Field

Consider an arbitrary optical electric field which is not a pure harmonic wave, but consists of components of more or less different frequencies. We now express such an optical electric field $E(t)$, which varies with time, by the complex Fourier integral

$$E(t) = \int_{-\infty}^{\infty} f(\omega)\, e^{i\omega t}\, d\omega \,, \tag{2.10}$$

where $f(\omega)$ is the complex Fourier component of frequency ω given by the complex Fourier transform

$$f(\omega) = \frac{1}{2\pi} \int_{-\infty}^{\infty} E(t)\, e^{-i\omega t}\, dt \,. \tag{2.11}$$

Since an actual electric field is real, $E(t)$ must be real. This is so if the condition

$$f(-\omega) = f(\omega)^* \tag{2.12}$$

is fulfilled, where the asterisk (*) denotes the complex conjugate. Therefore, if the positive frequency component $f(\omega)$ is known, the negative frequency component can be determined uniquely from (2.12).

Then, as an analytical expression for the optical electric field $E(t)$ at a point, the complex amplitude $A(t)$ consisting of only the positive frequency components was defined by *D. Gabor* in 1946 [2.2] in the form

$$A(t) = 2 \int_{0}^{\infty} f(\omega)\, e^{i\omega t}\, d\omega \,. \tag{2.13}$$

The complex amplitude, thus defined, is used to express the actual amplitude as

$$E(t) = \mathrm{Re}\,\{A(t)\} \,. \tag{2.14}$$

The imaginary part of (2.13) and its real part, (2.14), are mutual Hilbert transforms of one another and can be expressed as

$$\mathrm{Im}\,\{A(t)\} = -\frac{1}{\pi} \int_{-\infty}^{\infty} \frac{E(t')}{t'-t}\, dt' \,,$$

$$E(t) = \frac{1}{\pi} \int_{-\infty}^{\infty} \frac{\mathrm{Im}\,\{A(t')\}}{t'-t}\, dt' \,,$$

where the principal value is taken at $t = t'$ in the integral.

The complex representation of the electric field (or magnetic field) of light or radio waves is employed in the circuit theory of alternating currents and the electromagnetic theory of light to describe nearly perfectly monochromatic waves. By defining the complex amplitude analytically, as above, we are now able to deal with electromagnetic waves that change arbitrarily in time and are accordingly not monochromatic.

The light energy in a homogeneous medium is generally proportional to the square of the amplitude of the electric field. Therefore, the light intensity is given by the square of the amplitude. However, it is more appropriate to consider the square of the absolute value of the complex amplitude, namely $A(t)A^*(t)$, rather than the square of the actual electric field $E(t)$. For example, $[E(t)]^2$ for an optical pulse of frequency ω has a component of frequency 2ω, which must be removed, usually by averaging over a time longer than $1/\omega$. If we do this, the true instantaneous light intensity will be lost, because the envelope of the pulse certainly changes during the time over which the average is taken. But since $A(t)A^*(t)$ represents only the pulse envelope without involving components of optical frequencies, we are able to express the instantaneous light intensity. Thus we shall define the instantaneous intensity $I(t)$ whose amplitude is given by (2.13) as

$$I(t) = |A(t)|^2 = A(t)A^*(t) . \tag{2.15}$$

Here the proportionality constant is taken to be unity for light propagating in a vacuum or in a homogeneous medium. When describing light in an inhomogeneous medium, however, a proportionality constant determined by the refractive index of the medium must be included. It should be noted, in addition, that the equation is approximate when the refractive index is dispersive or anisotropic.

Let us now use the complex representation to calculate the light intensity entering the detector D in the Michelson interferometer. The light beams divided into two by the half-silvered mirror are superposed at D with equal amplitude after being reflected by the mirrors M_1 and M_2. Since there is an optical path difference $s_2 - s_1$, the light beam reflected by M_2 has a time lag $\tau = (s_2 - s_1)/c$. Thus, the light intensity entering the detector can be expressed as

$$\begin{aligned} I(t) &= [A(t) + A(t+\tau)]\,[A^*(t) + A^*(t+\tau)] \\ &= A(t)A^*(t) + A(t+\tau)A^*(t+\tau) \\ &\quad + A(t)A^*(t+\tau) + A(t+\tau)A^*(t) . \end{aligned}$$

We take the time average of this, since we observe only the time average of intensity in the experiment. For a light source of constant intensity, the intensity at times t and $t + \tau$ are equal, hence

$$\overline{A(t)A^*(t)} = \overline{A(t+\tau)A^*(t+\tau)} ,$$

both being equal to the constant $|A|^2$. Here the bars over the terms denote time averaging. Now, since

$$\overline{A(t)A^*(t+\tau)} = \overline{[A^*(t)A(t+\tau)]^*} ,$$

we may write

$$\overline{I(t)} = 2|A|^2 + 2\,\mathrm{Re}\,\overline{\{A^*(t)A(t+\tau)\}}\ . \tag{2.16}$$

The second term on the right-hand side of the above equation is equivalent to the real part of the autocorrelation function given by

$$\begin{aligned} G(\tau) &= \langle A^*(t)A(t+\tau)\rangle \\ &= \lim_{T\to\infty}\frac{1}{2T}\int_{-T}^{T} A^*(t)A(t+\tau)\,dt\ . \end{aligned} \tag{2.17}$$

The reason for this is because the ensemble average denoted by the brackets $\langle\ \rangle$ and the time average are equivalent to each other when the fluctuation is statistically stationary as for ordinary light. If we measure the variation of the fringe visibility of the interference pattern in the Michelson interferometer as a function of $\tau = (s_2 - s_1)/c$ by changing the distance l_2 to the mirror, we obtain the relative variation of the autocorrelation function as a function of the time difference τ.

The autocorrelation function $G(\tau)$ as given above is the Fourier transform of the power spectrum $I(\omega)$, that is, they are mutually related by the equations

$$G(\tau) = \int_{-\infty}^{\infty} I(\omega)\,\mathrm{e}^{\mathrm{i}\omega\tau}\,d\omega\ , \tag{2.18}$$

$$I(\omega) = \frac{1}{2\pi}\int_{-\infty}^{\infty} G(\tau)\,\mathrm{e}^{-\mathrm{i}\omega\tau}\,d\tau\ . \tag{2.19}$$

These equations constitute the Wiener-Khintchine theorem.

2.5 Coherence Function

Although the autocorrelation function represents temporal coherence, in order to make it more general and include spatial coherence as well, we shall consider the mutual correlation of light amplitudes between two points. If $A_1(t)$ and $A_2(t)$ are the complex amplitudes of light at two separate points P_1 and P_2, respectively, the mutual correlation function is given by

$$\begin{aligned} G_{12}(\tau) &= \langle A_1^*(t)A_2(t+\tau)\rangle \\ &= \lim_{T\to\infty}\frac{1}{2T}\int_{-T}^{T} A_1^*(t)A_2(t+\tau)\,dt\ . \end{aligned} \tag{2.20}$$

Then $G_{12}(\tau)$ just represents the light intensity on the screen in Young's experiment where the time difference is τ between the two components of light from the slits P_1 and P_2.

When the mutual correlation function $G_{12}(\tau)$ is normalized, it can be expressed as

$$\gamma_{12}(\tau) = \frac{G_{12}(\tau)}{\sqrt{G_{11}(0)\, G_{22}(0)}} \; . \tag{2.21}$$

F. Zernike, in 1938 [2.3], called this the first-order coherence function. Since $G_{11}(0) = \langle |A_1(t)|^2 \rangle$ represents the light intensity I_1 at P_1, and $G_{22}(0) = \langle |A_2(t)|^2 \rangle$ the light intensity I_2 at P_2, we may also write

$$\gamma_{12}(\tau) = \frac{\langle A_1^*(t)\, A_2(t+\tau) \rangle}{\sqrt{I_1 I_2}} \; . \tag{2.22}$$

$\gamma_{12}(\tau)$ is now called the complex degree of coherence. If the light waves at P_1 and P_2 are both perfectly spatially coherent and temporally coherent for the time difference τ, we have $|\gamma_{12}(\tau)| = 1$; if they are either spatially or temporally perfectly incoherent, we have $\gamma_{12}(\tau) = 0$.

In general, the light intensity when light waves $A_1(t)$ and $A_2(t+\tau)$ from two arbitrary points are superposed is given by

$$|A_1(t)|^2 + |A_2(t+\tau)|^2 + A_1(t)\, A_2^*(t+\tau) + A_1^*(t)\, A_2(t+\tau) \; .$$

Its time average, denoted by I, is obtained using (2.20, 22):

$$\begin{aligned} I &= I_1 + I_2 + G_{12}^*(\tau) + G_{12}(\tau) \\ &= I_1 + I_2 + 2\sqrt{I_1 I_2}\, \mathrm{Re}\, \{\gamma_{12}(\tau)\} \; . \end{aligned} \tag{2.23}$$

The complex degree of coherence $\gamma_{12}(\tau)$ becomes smaller, the further apart P_1 and P_2 are and the longer the time difference τ is. As long as the change in the path difference between light waves from P_1 and P_2 to the superposed point is confined to within a few wavelengths, the magnitude of $\gamma_{12}(\tau)$ does not effectively change for an ordinary monochromatic light source of fairly good coherence, but the phase of $\gamma_{12}(\tau)$ changes appreciably. Thus, we may write

$$\gamma_{12}(\tau) = \gamma_0 \exp[ik(s_2 - s_1)] \tag{2.24}$$

where the complex quantity γ_0 will be constant for a small change in $\tau = (s_2 - s_1)/c$. Then the intensity of the interference pattern at the bright and dark parts is given by

$$\begin{aligned} I_{\max} &= I_1 + I_2 + 2\sqrt{I_1 I_2}\, |\gamma_0| \; , \\ I_{\min} &= I_1 + I_2 - 2\sqrt{I_1 I_2}\, |\gamma_0| \; , \end{aligned}$$

and the visibility of the interference fringes from (2.8) can be expressed as

$$V = \frac{2\sqrt{I_1 I_2}}{I_1 + I_2}\, |\gamma_0| \; . \tag{2.25}$$

If, therefore, we measure the relative intensities of I_1 and I_2, and the visibility of the interference fringes, we can obtain the value of $|\gamma_0|$, that is, the magnitude of $\gamma_{12}(\tau)$. The phase of $\gamma_{12}(\tau)$ is 0 at the brightest part of the interference pattern and π at the darkest part.

Finally, as an important example, let us consider the coherence function of light whose frequency, although nearly monochromatic, is distributed over a narrow band as explained in Sect. 2.3. The normalized self-coherence function of light at P_1 is expressed as

$$\gamma_{11}(\tau) = \frac{\langle A_1^*(t) A_1(t+\tau)\rangle}{\langle A_1^*(t) A_1(t)\rangle} . \tag{2.26}$$

Since each atom in a light source emits a damped light wave as expressed by (2.9), the complex amplitude of the light wave emitted from the nth atom can be written

$$A_n(t) = A \exp(\mathrm{i}\omega_n t + i\theta_n - t/2\tau_a) . \tag{2.27}$$

If the frequency of each atom is the same, so that $\omega_n = \omega_0$, while the phase of emission θ_n is random, the self-coherence function becomes

$$\gamma_{11}(\tau) = \mathrm{e}^{\mathrm{i}\omega_0\tau}\, \mathrm{e}^{-|\tau|/2\tau_a} . \tag{2.28}$$

This is the same for all the light waves emitted by a large number of atoms in the light source. In this case the self-coherence function varies as shown in Fig. 2.7 and represents the damping of each atom.

Next, let us examine the case where the emission frequency distribution is Gaussian, while the lifetime of the excited atom is relatively long. We assume that the frequency distribution is centered at ω_0 and given by

$$I(\omega) = \frac{1}{\sqrt{\pi}\, w} \exp[-(\omega - \omega_0)^2/w^2] \tag{2.29}$$

where w is the halfwidth at the point where $I(\omega)$ becomes 1/e of $I(\omega_0)$. Since (2.29) is the frequency distribution of the light intensity and gives the power spectrum, the autocorrelation function can be obtained from (2.18):

$$\begin{aligned} \gamma_{11}(\tau) &= \frac{1}{\sqrt{\pi}\, w} \int_{-\infty}^{\infty} \exp\left[-\left(\frac{\omega - \omega_0}{w}\right)^2 + \mathrm{i}\omega\tau\right] d\omega \\ &= \mathrm{e}^{\mathrm{i}\omega_0\tau} \cdot \mathrm{e}^{-w^2\tau^2/4} . \end{aligned} \tag{2.30}$$

The autocorrelation function in this case is Gaussian, as shown in Fig. 2.8. The degree of coherence is seen to decrease when the time difference τ is larger than about $1/w$, even though the atomic damping time is much longer.

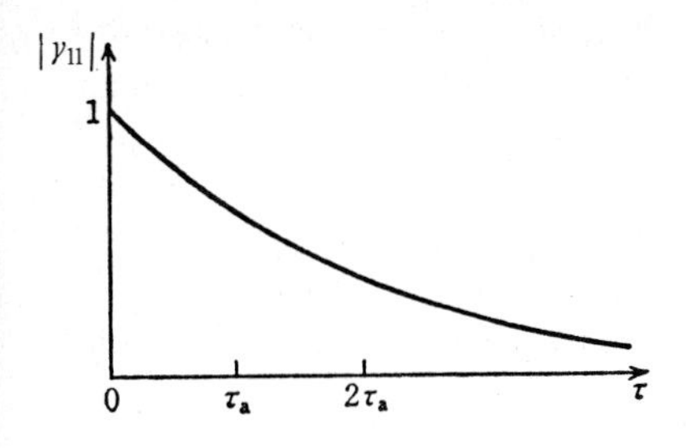

Fig. 2.7. The autocorrelation function given by (2.28)

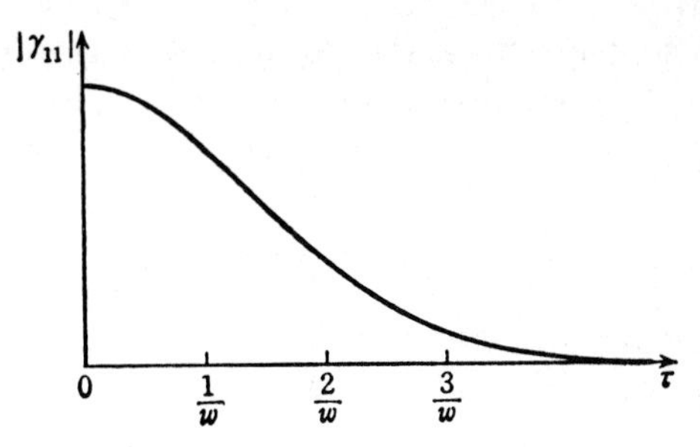

Fig. 2.8. The autocorrelation function given by (2.30)

Problems

2.1 In performing Young's experiment using a red He-Ne laser, we wish to observe interference fringes of 3 mm period on a screen at a distance 1 m from the double slit. Find the separation of the slits in this experiment.

Answer: 0.21 mm.

2.2 Explain why the slit P_0 in Fig. 2.2 is necessary in Young's experiment. What happens if the slit P_0 is too wide or too narrow?

2.3 Calculate the visibility of the interference fringes as a function of the path difference in an ideal Michelson interferometer for an ensemble of damped oscillations of light, as given by (2.9).

Answer: $V = \exp(-\Delta s/2c\tau_a)$.

2.4 Suppose that the intensity of interference fringes observed by a Michelson interferometer is expressed by

$$I(\Delta s) = I_0(1 + e^{-a^2(\Delta s)^2} \cos 2k\Delta s) ,$$

where $\Delta s = s_2 - s_1$ is the optical path difference and the constant a is much smaller than the wavenumber k. Find the spectrum (power spectral density) of the light source.

Hint: Use (2.30).
Answer: $I(\omega) = (1/\sqrt{\pi}\, ac) \exp[-(\omega - \omega_0)^2/a^2c^2]$.

2.5 Verify that $\gamma_{12}(\tau) = \gamma_{21}^*(-\tau)$, where $\gamma_{12}(\tau)$ is the complex degree of coherence or the normalized mutual correlation function.

2.6 Calculate the power spectrum for a light source for which the self-coherence function is given by (2.28).

Answer: $I(\omega) = \{2\pi\tau_a[(\omega - \omega_0)^2 + 1/4\,\tau_a^2]\}^{-1}$

2.7 The complex amplitude of a Gaussian pulse of light with a linear frequency chirp is written as

$$A(t) = A_0 \exp\left(-\frac{t^2}{\tau_p^2} + i\omega_0 t + iat^2\right) .$$

Find the power spectrum and the normalized self-coherence function of this light pulse.

Hint: $\int_{-\infty}^{\infty} \exp[-(x + iy)^2]\, dx = \sqrt{\pi}$.

Answer: $I(\omega) = [\pi(1 + a^2\tau_p^2)]^{-1/2} \exp[-(\omega - \omega_0)^2 (1 + a^2\tau_p^2)^{-1}\, \tau_p^2/2]$,

$\gamma(\tau) = \exp[-i\omega_0\tau - (1 + a^2)\,\tau^2/2]$.

2.8 Suppose that a spectral line has a Gaussian profile with a full width at half maximum of 1 GHz. What is the optical path difference at which the visibility falls to 0.5 in an ideal interferometer when using this light source?

Answer: $\tau_{1/2} = 4 \ln 2/\mathrm{FWHM}$, $\Delta s_{1/2} = 83.1$ cm.

3. Electromagnetic Theory of Light

The wave nature of light is copiously manifest in experiments on interference and diffraction, and it is theoretically depicted as electromagnetic waves, according to Maxwell's equations. In contrast, the particle nature of light is expressed through the idea of a light quantum or a photon resulting from quantization of the electromagnetic field. However, the coherence of laser light is much better than that of other forms of light, and it is only in exceptional cases that the quantization of the electromagnetic field of laser light manifests any substantial effect. Therefore, in this chapter, we shall discuss the nature of light as expressed in terms of classical electromagnetic waves.

3.1 Maxwell's Equations

As is known from electromagnetic theory, the electric field $\boldsymbol{E}$, electric flux density $\boldsymbol{D}$, magnetic field $\boldsymbol{H}$, magnetic flux density $\boldsymbol{B}$, electric current density $\boldsymbol{j}$ and charge density ϱ, all of which may change as functions of the coordinates (x, y, z) and time t, are related by Maxwell's equations:

$$\operatorname{curl} \boldsymbol{E} = -\frac{\partial \boldsymbol{B}}{\partial t}, \tag{3.1}$$

$$\operatorname{curl} \boldsymbol{H} = \boldsymbol{j} + \frac{\partial \boldsymbol{D}}{\partial t}, \tag{3.2}$$

$$\operatorname{div} \boldsymbol{D} = \varrho, \tag{3.3}$$

$$\operatorname{div} \boldsymbol{B} = 0. \tag{3.4}$$

Sometimes $\operatorname{curl} \boldsymbol{E}$ is written as $\operatorname{rot} \boldsymbol{E}$ or $\nabla \times \boldsymbol{E}$, and $\operatorname{div} \boldsymbol{D}$ as $\nabla \cdot \boldsymbol{D}$. Here ∇ is the vector operator with $\partial/\partial x$, $\partial/\partial y$, and $\partial/\partial z$ as its x, y, and z components, respectively. By using this operator, $\operatorname{grad} \phi$ can be expressed as $\nabla\phi$.

Let ε denote the electric permittivity (dielectric constant), μ the magnetic permeability and σ the electric conductivity of the medium; then we have

$$\boldsymbol{D} = \varepsilon \boldsymbol{E}, \quad \boldsymbol{B} = \mu \boldsymbol{H}, \quad \boldsymbol{j} = \sigma \boldsymbol{E}. \tag{3.5}$$

By using the polarization $\boldsymbol{P}$ and the permittivity in vacuo ε_0, we have

$$\boldsymbol{D} = \varepsilon_0 \boldsymbol{E} + \boldsymbol{P},$$

$$\varepsilon_0 = 8.854 \times 10^{-12}\,\mathrm{F/m}.$$

The quantity given by

$$\chi = \frac{\boldsymbol{P}}{\varepsilon_0 \boldsymbol{E}} \tag{3.6}$$

is called the electric susceptibility, and we have $\varepsilon = \varepsilon_0(1 + \chi)$. In general, $\boldsymbol{P}$ is proportional to $\boldsymbol{E}$ when the electric field $\boldsymbol{E}$ is weak, but it is no longer proportional when $\boldsymbol{E}$ is strong. Moreover, it does not always follow that $\boldsymbol{P}$ varies in time in accordance with the time variation of $\boldsymbol{E}$. Cases where such nonlinear effects or dispersion exist will be treated in later chapters. Here, both χ and ε are assumed to be constants. We may also assume that the medium is dielectric such that $\sigma = 0$ and the permeability is

$$\mu = \mu_0 = 4\pi \times 10^{-7}\,\mathrm{H/m}\ .$$

Now, taking the curl of (3.1) and using (3.2) and $\boldsymbol{B} = \mu\boldsymbol{H}$, curl curl $\boldsymbol{E}$ becomes

$$\nabla \times \nabla \times \boldsymbol{E} = -\mu\frac{\partial^2 \boldsymbol{D}}{\partial t^2}\ . \tag{3.7}$$

According to vector calculation we have

$$\nabla \times \nabla \times \boldsymbol{E} = -\nabla^2 \boldsymbol{E} + \nabla(\nabla \cdot \boldsymbol{E})\ .$$

Using (3.3) and $\boldsymbol{D} = \varepsilon\boldsymbol{E}$, we have

$$\nabla \cdot \boldsymbol{E} = \frac{1}{\varepsilon}\nabla \cdot \boldsymbol{D} = \frac{\varrho}{\varepsilon}\ .$$

But, since an electrostatic charge produces only an electrostatic field and is irrelevant to electromagnetic waves in an optical medium, we shall neglect it and put $\varrho = 0$. Therefore, (3.7) becomes

$$\nabla^2 \boldsymbol{E} - \varepsilon\mu\frac{\partial^2 \boldsymbol{E}}{\partial t^2} = 0\ , \tag{3.8a}$$

namely,

$$\left(\frac{\partial^2}{\partial x^2} + \frac{\partial^2}{\partial y^2} + \frac{\partial^2}{\partial z^2} - \varepsilon\mu\frac{\partial^2}{\partial t^2}\right)\boldsymbol{E} = 0\ . \tag{3.8b}$$

This is the equation of waves propagating with the velocity

$$v = \frac{1}{\sqrt{\varepsilon\mu}}\ . \tag{3.9}$$

The velocity c of light in vacuo is

$$c = \frac{1}{\sqrt{\varepsilon_0\mu_0}} = 299\ 792\ 458\ \mathrm{m/s}\ .$$

Denoting the refractive index of the medium by η, we have

$$\eta = \sqrt{\frac{\varepsilon\mu}{\varepsilon_0\mu_0}} = c\sqrt{\varepsilon\mu}\,, \quad v = \frac{c}{\eta}\,. \tag{3.10}$$

For an ideal plane wave, not only is the phase the same everywhere on its planar wave surface but the magnitude and direction of the electric field are also the same everywhere. If the wave surface is parallel to the xy plane, the derivatives of $\boldsymbol{E}$ with respect to x and y are both zero, so that (3.8 b) becomes

$$\frac{\partial^2 \boldsymbol{E}}{\partial z^2} - \frac{1}{v^2}\frac{\partial^2 \boldsymbol{E}}{\partial t^2} = 0\,. \tag{3.11}$$

This is the equation of waves propagating in the $\pm z$ directions, and its solution can be expressed, generally, in the form

$$E_x = f_1(z - vt) + f_2(z + vt)\,, \tag{3.12 a}$$

$$E_y = g_1(z - vt) + g_2(z + vt)\,. \tag{3.12 b}$$

The electromagnetic waves are transverse waves from Maxwell's equations, so that here we have $E_z = 0$. The quantities f_1, f_2, g_1, and g_2 in (3.12) are single-valued functions. The function $f_1(z)$ gives the waveform, at $t = 0$, of the component of the plane wave polarized in the x direction and propagating in the $+z$ direction, while $f_2(z)$ gives the waveform of the plane wave propagating in the $-z$ direction. Similarly, $g_1(z)$ and $g_2(z)$ give the waveforms of the plane waves polarized in the y direction and propagating in the $\pm z$ directions, respectively.

From (3.2) the waves of the magnetic field accompanying the propagation of the electric field are given as follows. The x component of (3.2) becomes

$$-\frac{\partial H_y}{\partial z} = \varepsilon\frac{\partial}{\partial t}E_x$$

because the derivatives with respect to x and y are both zero for a plane wave propagating in the z direction. Now, differentiating (3.12 a) with respect to t and integrating with respect to z, we obtain

$$H_y = \varepsilon v\,[f_1(z - vt) - f_2(z + vt)]\,. \tag{3.13 a}$$

Similarly, from the y component of (3.2), we obtain

$$H_x = \varepsilon v\,[-g_1(z - vt) + g_2(z + vt)]\,. \tag{3.13 b}$$

Equations (3.12, 13) show that the fields $\boldsymbol{E}$ and $\boldsymbol{H}$ of each electromagnetic wave, as given by the functions f_1, f_2, g_1, and g_2, are perpendicular and propagate in the direction of the vector $\boldsymbol{E} \times \boldsymbol{H}$.

The electromagnetic energy flowing through a unit area normal to the z axis per unit time is given by the time average of the z component of the Poynting vector $\boldsymbol{S} = \boldsymbol{E} \times \boldsymbol{H}$. For a wave polarized in the x direction, we have

$$S_z = \varepsilon v [f_1^2(z - vt) - f_2^2(z + vt)]$$

from (3.12a, 13a). For a wave polarized in the y direction we have

$$S_z = \varepsilon v [g_1^2(z - vt) - g_2^2(z + vt)]$$

from (3.12b, 13b).

So far we have taken the electromagnetic wave to be of arbitrary form. Now, any waveform can be expressed as a superposition of harmonic waves using a Fourier expansion. On the other hand, laser light is almost perfectly monochromatic. If, therefore, we express the time factor of a monochromatic electromagnetic wave of frequency ω by $\exp(\mathrm{i}\omega t)$[1], the wave equation (3.8a) becomes

$$\nabla^2 \boldsymbol{E} + k^2 \boldsymbol{E} = 0 , \tag{3.14}$$

where $k^2 = \omega^2 \varepsilon \mu$. Here

$$k = \frac{\omega}{v} = \frac{2\pi}{\lambda}$$

is called the wavenumber or the phase constant, but when k is a complex quantity it is generally called the propagation constant.

Now the electric and magnetic fields of monochromatic light propagating in the $\pm z$ directions and polarized in the x direction can be expressed as

$$\begin{aligned} E_x &= F_1 \mathrm{e}^{\mathrm{i}\omega t - \mathrm{i}kz} + F_2 \mathrm{e}^{\mathrm{i}\omega t + \mathrm{i}kz} , \\ H_y &= \varepsilon v (F_1 \mathrm{e}^{\mathrm{i}\omega t - \mathrm{i}kz} - F_2 \mathrm{e}^{\mathrm{i}\omega t + \mathrm{i}kz}) . \end{aligned} \tag{3.15}$$

The first term is the component propagating in the $+z$ direction and the second term is the component propagating in the $-z$ direction. From (3.10) εv can be written, when $\mu = \mu_0$, as

$$\varepsilon v = \eta \varepsilon_0 c = \eta \sqrt{\frac{\varepsilon_0}{\mu_0}} . \tag{3.16}$$

For a plane electromagnetic wave propagating in an arbitrary direction we take a wave vector $\boldsymbol{k}$ in the direction of propagation, its magnitude being $k = \omega/v$. Then the complex electric field $\boldsymbol{E}(r, t)$ at a point specified by $\boldsymbol{r}$ is given by

$$\boldsymbol{E}(\boldsymbol{r}, t) = \boldsymbol{E}_0 \exp(\mathrm{i}\omega t - \mathrm{i}\boldsymbol{k} \cdot \boldsymbol{r}) \tag{3.17}$$

where $\boldsymbol{E}_0$ is perpendicular to $\boldsymbol{k}$ and denotes the complex amplitude of the plane wave. If the angles between the wave vector and the x, y, and z axes

---

1 The time factor is sometimes written $\exp(-\mathrm{i}\omega t)$, in which case the complex permittivity and the coefficients of reflection and transmission all become complex conjugates. It is customary to use $\exp(\mathrm{i}\omega t)$ in electronics and $\exp(-\mathrm{i}\omega t)$ in optics.

are taken to be α, β, and γ, respectively, the components of the wave vector are

$$k_x = k\cos\alpha, \quad k_y = k\cos\beta, \quad k_z = k\cos\gamma\ ,$$

and (3.17) is rewritten

$$\boldsymbol{E}(x, y, z, t) = \boldsymbol{E}_0 \exp[\mathrm{i}\omega t - \mathrm{i}k(x\cos\alpha + y\cos\beta + z\cos\gamma)]\ .$$

3.2 Reflection and Refraction of Light

When a plane electromagnetic wave is incident on a boundary plane between two media of different refractive indices, the reflection and transmission coefficients of each polarization component are determined by the conditions on the electromagnetic fields at the boundary. Here we take the boundary to be the xy plane, with the refractive index $\eta_1 = \sqrt{\varepsilon_1/\varepsilon_0}$ for $z < 0$, and $\eta_2 = \sqrt{\varepsilon_2/\varepsilon_0}$ for $z > 0$; the plane of incidence is taken as the xz plane and the incident angle as θ.

Since both the reflection and transmission coefficients depend on the polarization of the incident light, we shall treat the two components separately: one is the p-component whose electric field is parallel to the plane of incidence and the other is the s-component whose electric field is perpendicular to the plane of incidence[2]. If the positive direction of the electric field of the p-component is taken as in Fig. 3.1, the electric field and the magnetic field of the p-component of the incident light can be written as

$$\begin{aligned} E_\mathrm{p} &= A_\mathrm{p} \exp[-\mathrm{i}k_1(x\sin\theta + z\cos\theta)]\ , \\ H_y &= \varepsilon_1 v_1 E_\mathrm{p}\ , \quad H_x = H_z = 0 \end{aligned} \tag{3.18}$$

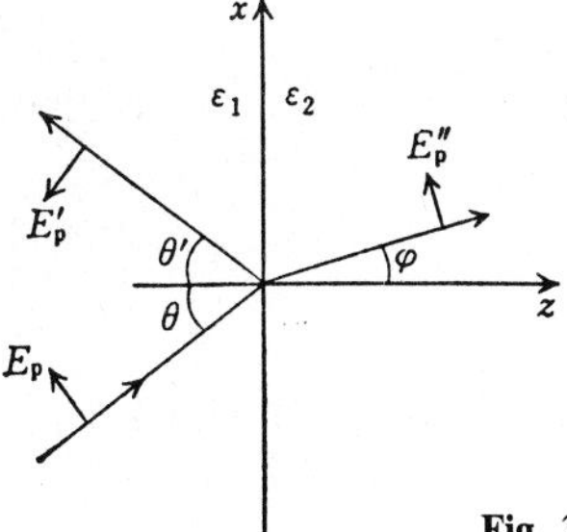

Fig. 3.1. Reflection and refraction of light

2 s stands for the German word *senkrecht*, meaning perpendicular.

where the time factor $\exp(i\omega t)$ has been dropped. The electric field of the s-component is in the y direction and is given by

$$E_s = A_s \exp[-ik_1(x \sin\theta + z\cos\theta)] . \tag{3.19}$$

The magnetic field of the s-component is perpendicular to both $\boldsymbol{k}$ and the y axis, and its magnitude is $\varepsilon_1 v_1 E_s$.

First, let us consider the p-component. Denoting the quantities pertaining to reflected light by a prime and taking the angle of reflection to be θ' as in Fig. 3.1, we can write the electric and magnetic fields of the reflected light as

$$\begin{aligned} E'_p &= A'_p \exp[-ik_1(x\sin\theta' - z\cos\theta')] , \\ H'_y &= \varepsilon_1 v_1 E'_p . \end{aligned} \tag{3.20}$$

The transmitted light for $z > 0$ with the angle of refraction φ is expressed by

$$\begin{aligned} E''_p &= A''_p \exp[-ik_2(x\sin\varphi + z\cos\varphi)] , \\ H''_y &= \varepsilon_2 v_2 E''_p . \end{aligned} \tag{3.21}$$

The boundary conditions at the boundary plane $z = 0$ require that the x- and y-components of $\boldsymbol{E}$ and the z-component of $\boldsymbol{D} = \varepsilon\boldsymbol{E}$ be continuous, while all the components of $\boldsymbol{H}$ are continuous (assuming $\mu_1 = \mu_2 = \mu_0$). The electric and magnetic fields of the p-component in the medium on the side $z < 0$ are, as $z \rightarrow 0$,

$$\begin{aligned} E_x &= A_p \cos\theta\, e^{-ik_1 x \sin\theta} - A'_p \cos\theta'\, e^{-ik_1 x\sin\theta'} , \\ H_y &= \varepsilon_1 v_1 (A_p e^{-ik_1 x\sin\theta} + A'_p\, e^{-ik_1 x\sin\theta'}) . \end{aligned}$$

These must be respectively equal to the values of the electric and magnetic fields for $z > 0$, as $z \rightarrow 0$. Then we have

$$A_p \cos\theta\, e^{-ik_1 x\sin\theta} - A'_p \cos\theta'\, e^{-ik_1 x\sin\theta'} = A''_p \cos\varphi\, e^{-ik_2 x\sin\varphi} , \tag{3.22}$$

$$\eta_1 (A_p e^{-ik_1 x\sin\theta} + A'_p e^{-ik_1 x\sin\theta'}) = \eta_2 A''_p e^{-ik_2 x\sin\varphi} , \tag{3.23}$$

where (3.16) has been used to obtain (3.23). Since (3.22, 23) hold at any point with an arbitrary value of x on the boundary plane, we have

$$k_1 \sin\theta = k_1 \sin\theta' = k_2 \sin\varphi . \tag{3.24}$$

From this we obtain the law of reflection

$$\theta = \theta'$$

and the law of refraction

$$\frac{\sin\theta}{\sin\varphi} = \frac{k_2}{k_1} = \frac{\eta_2}{\eta_1} . \tag{3.25}$$

By using (3.24), (3.22, 23) are reduced to

$$(A_p - A'_p)\cos\theta = A''_p\cos\varphi\ ,$$
$$\eta_1(A_p + A'_p) = \eta_2 A''_p\ . \tag{3.26}$$

Eliminating A'' from these equations, we obtain the amplitude reflection coefficient:

$$r_p = \frac{A'_p}{A_p} = \frac{\sin\theta\cos\theta - \sin\varphi\cos\varphi}{\sin\theta\cos\theta + \sin\varphi\cos\varphi}\ . \tag{3.27}$$

Rewriting this, the amplitude and power reflection coefficients become, respectively,

$$r_p = \frac{\tan(\theta-\varphi)}{\tan(\theta+\varphi)}\ ,\qquad R_p = \frac{\tan^2(\theta-\varphi)}{\tan^2(\theta+\varphi)}\ . \tag{3.28}$$

The power reflection coefficient R_p is often called the reflectance.

We see that from (3.27 or 28), the reflection coefficient of the p-component is 0 when $\theta + \varphi = \pi/2$. This value of the incident angle is called the Brewster angle, and is denoted by θ_B, given by

$$\tan\theta_B = \frac{\eta_2}{\eta_1}\ . \tag{3.29}$$

With $\eta_1 = 1$ we have $\theta_B = 56°\,40'$ for $\eta_2 = 1.52$ (BK7 glass), $\theta_B = 55°\,35'$ for $\eta_2 = 1.46$ (quartz glass) and $\theta_B = 75°\,58'$ for $\eta_2 = 4.0$ (germanium at wavelengths 1.5–10 μm).

By substituting (3.27) into either one of (3.26), we obtain the amplitude transmission coefficient of the p-component:

$$d_p = \frac{A''_p}{A_p} = \frac{2\cos\theta\sin\varphi}{\sin\theta\cos\theta + \sin\varphi\cos\varphi} \quad \text{or}$$
$$d_p = \frac{2\cos\theta\sin\varphi}{\sin(\theta+\varphi)\cos(\theta-\varphi)}\ . \tag{3.30}$$

Since this is the ratio of the amplitudes of the electric fields in two different media, d_p^2 does not represent the transmission coefficient of the light intensity.

If we define the light intensity by the power density, as given by the energy passing through a unit area normal to the wave vector in a unit time, it is equal to $\boldsymbol{E} \times \boldsymbol{H}$. Consequently, since $\boldsymbol{E} \times \boldsymbol{H}$ is proportional to $\varepsilon v E^2$ or ηE^2, the power transmission coefficient or the transmittance of the p-component becomes

$$D_p = \frac{\eta_2 A''^2_p}{\eta_1 A^2_p} = \frac{4\sin\theta\cos^2\theta\sin\varphi}{\sin^2(\theta+\varphi)\cos^2(\theta-\varphi)}\ . \tag{3.31}$$

However, it should be noticed that when we consider the transmission coefficient of the energy flux of a light beam, it is

$$T_p = \frac{\sin 2\theta \sin 2\varphi}{\sin^2(\theta+\varphi)\cos^2(\theta-\varphi)} \tag{3.32}$$

since the cross-sectional area of the light beam changes due to refraction by the factor

$$\frac{\cos\varphi}{\cos\theta} .$$

Similarly, the amplitude reflection and transmission coefficients for the s-component are, respectively,

$$r_s = \frac{A_s'}{A_s} = -\frac{\sin(\theta-\varphi)}{\sin(\theta+\varphi)} \quad \text{and} \tag{3.33}$$

$$d_s = \frac{A_s''}{A_s} = \frac{2\cos\theta\sin\varphi}{\sin(\theta+\varphi)} . \tag{3.34}$$

The power transmission coefficients, or transmittance, and the transmission coefficient of the energy flux are, respectively,

$$D_s = \frac{4\sin\theta\cos^2\theta\sin\varphi}{\sin^2(\theta+\varphi)} \quad \text{and} \tag{3.35}$$

$$T_s = \frac{\sin 2\theta \sin 2\varphi}{\sin^2(\theta+\varphi)} . \tag{3.36}$$

These equations, giving the reflection and transmission coefficients at the boundary of two different media, are known as the Fresnel formulas.

3.3 Total Reflection

In Fig. 3.1 rays of light are shown for the case when $\eta_2 > \eta_1$. When $\eta_2 < \eta_1$, however, there is an angle of incidence θ_c which gives an angle of refraction $\varphi = \pi/2$. If the angle of incidence θ is greater than θ_c, there will be no transmitted ray. This angle θ_c is called the critical angle and is given by

$$\sin\theta_c = \frac{1}{n_{12}} , \quad n_{12} = \frac{\eta_1}{\eta_2} > 1 . \tag{3.37}$$

Since all the incident light is reflected when $\theta > \theta_c$, this situation is called total reflection, which means that there will be no plane wave (3.21) propagating at large distances from the boundary; it does not, however, mean that the electromagnetic fields vanish completely for $z > 0$.

Although there is no real angle of refraction that satisfies (3.25) in total reflection when $\theta > \theta_c$, a complex value of $\varphi = ia + \pi/2$ will formally satisfy the law of refraction (3.25), when

$$\sin\varphi = n_{12}\sin\theta > 1\,. \tag{3.38}$$

Thus, if we take $\cos\varphi$ to be the imaginary quantity

$$\cos\varphi = \pm\sqrt{1-\sin^2\varphi} = \pm\mathrm{i}\sqrt{n_{12}^2\sin^2\theta - 1}\,,$$

then all the Fresnel equations given in Sect. 3.2 will hold. The expression for the p- or s-component of the refracted wave becomes

$$E'' = A''\exp[-ik_2(xn_{12}\sin\theta \pm \mathrm{i}z\sqrt{n_{12}^2\sin^2\theta - 1})]\,.$$

Since the amplitude of the p- and s-components must be 0 as $z \to \infty$, we take the negative sign in front of the square root in $\cos\varphi$, namely,

$$\cos\varphi = -\mathrm{i}\sqrt{n_{12}^2\sin^2\theta - 1}\,. \tag{3.39}$$

Thus the electric field in the medium for $z > 0$ is

$$E'' = A''\exp(-k_2\sqrt{n_{12}^2\sin^2\theta - 1}\,z)\exp[-\mathrm{i}(k_2 n_{12}\sin\theta)x]\,. \tag{3.40}$$

Since the factor $\exp(\mathrm{i}\omega t)$ has been dropped, this equation shows that the amplitude of the plane wave propagating in the x direction diminishes exponentially with z. Such a wave is called an evanescent wave.

The depth z_{ev} at which the amplitude of the evanescent wave becomes 1/e of its amplitude at $z = 0$ is given by

$$z_{ev} = \frac{1}{k_2\sqrt{n_{12}^2\sin^2\theta - 1}}\,. \tag{3.41}$$

If θ is near the critical angle θ_c, we may approximate (3.41) for a small value of $\Delta\theta = \theta - \theta_c$ by

$$z_{ev} \simeq \frac{\lambda_0}{2\pi\eta_2}\sqrt{\frac{\tan\theta_c}{2\Delta\theta}}$$

where λ_0 is the wavelength in vacuo. It can be seen from this that the penetration depth of the evanescent wave is about one wavelength when $\Delta\theta \cong 0.01$ rad.

By substituting (3.38, 39) into (3.30), we obtain from (3.40)

$$E''_p = \frac{2n_{12}\cos\theta}{\cos\theta - \mathrm{i}n_{12}\sqrt{n_{12}^2\sin^2\theta - 1}} \times A_p\exp(-z/z_{ev})\exp[-\mathrm{i}(k_2 n_{12}\sin\theta)x]\,. \tag{3.42}$$

Then the electric and magnetic fields of the evanescent wave due to the p-component of the incident wave are given by

$$E_x = -\mathrm{i}\sqrt{n_{12}^2 \sin^2\theta - 1}\, E_p'' \,, \tag{3.43}$$

$$E_z = -n_{12} \sin\theta E_p'' \,, \tag{3.44}$$

$$H_y = \varepsilon_2 v_2 E_p'' \,. \tag{3.45}$$

The other components, E_y, H_x, and H_z, are all zero.

The z component of the Poynting vector is

$$S_z = E_x H_y - E_y H_x = E_x H_y \,.$$

It is clear from (3.43, 45) that E_x and H_y differ in phase by 90°, so that the time average of S_z is 0 in this case. On the other hand, it can be seen from (3.44, 45) that, for $\sin\theta > 0$, E_z and H_y have opposite phases, so that the time average of the x component of the Poynting vector given by

$$S_x = E_y H_z - E_z H_y = -E_z H_y$$

takes a positive value. The energy of the evanescent wave flows along the x direction in the plane of incidence near the boundary plane. This may be thought of as being due to the incident light not being totally reflected immediately at the geometrical boundary $z = 0$, but being reflected after having penetrated a small depth z_{ev} into the medium at $z > 0$. In fact, when a pencil beam of light is totally reflected, there exists a certain shift between the positions of the incident and the reflected beams at the surface as shown in Fig. 3.2. This shift is called the Goos-Hänchen shift. It becomes large as θ approaches the critical angle θ_c, but even then it is only of the order of the wavelength [3.1].

If two totally reflecting prisms are close together as in Fig. 3.3, so that their separation is of the order of one wavelength, the evanescent wave developed by total reflection at the upper prism penetrates into the lower

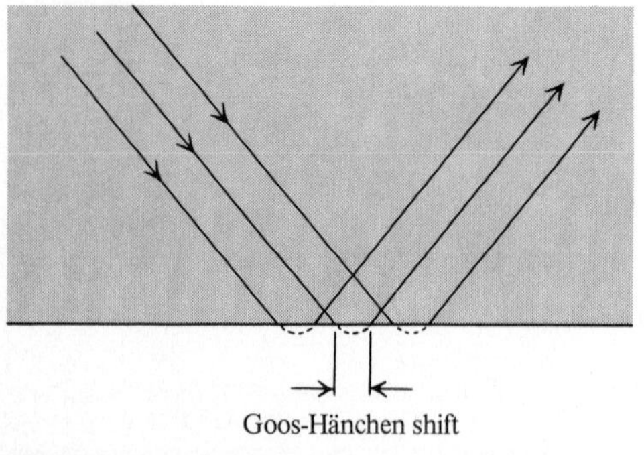

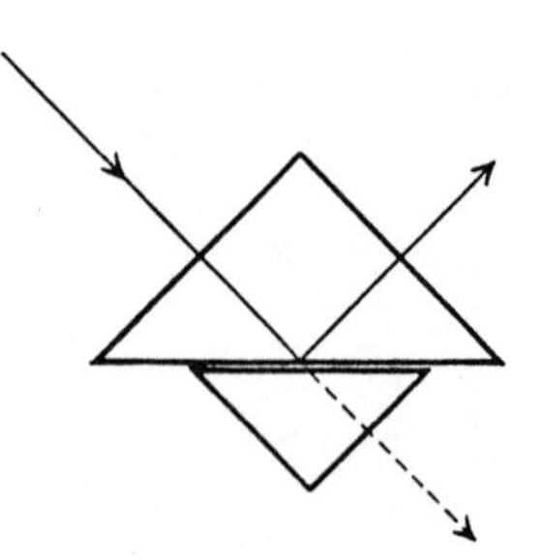

Fig. 3.2. Goos-Hänchen shift at total reflection
Fig. 3.3. Prism coupling with an evanescent wave ▶

prism, resulting in partial transmission of light into the lower prism, as shown by the dotted line. The extent of coupling varies appreciably with the separation of the two prisms. A device based on this principle is used for coupling a light beam in free space with a guided optical wave [3.2].

Although the power reflection coefficient at total reflection is 1, the amplitude reflection coefficient is not equal to 1 owing to a shift in phase. From (3.27, 33) we obtain

$$r_{\mathrm{p}} = \frac{\cos\theta + \mathrm{i} n_{12}\sqrt{n_{12}^2 \sin^2\theta - 1}}{\cos\theta - \mathrm{i} n_{12}\sqrt{n_{12}^2 \sin^2\theta - 1}} ,$$

$$r_{\mathrm{s}} = \frac{n_{12}\cos\theta + \mathrm{i}\sqrt{n_{12}^2 \sin\theta - 1}}{n_{12}\cos\theta - \mathrm{i}\sqrt{n_{12}^2 \sin^2\theta - 1}} .$$

Since the absolute values of these coefficients are equal to 1, they can be rewritten as

$$r_{\mathrm{p}} = \mathrm{e}^{\mathrm{i}\delta_{\mathrm{p}}} , \quad r_{\mathrm{s}} = \mathrm{e}^{\mathrm{i}\delta_{\mathrm{s}}} , \tag{3.46}$$

where the phase angles δ_{p} and δ_{s} are given by

$$\tan\frac{\delta_{\mathrm{p}}}{2} = \frac{n_{12}\sqrt{n_{12}^2 \sin^2\theta - 1}}{\cos\theta} , \tag{3.47}$$

$$\tan\frac{\delta_{\mathrm{s}}}{2} = \frac{\sqrt{n_{12}^2 \sin^2\theta - 1}}{n_{12}\cos\theta} . \tag{3.48}$$

The variation of δ_{p} and δ_{s} with θ is shown in Fig. 3.4 for $n_{12} = 1.52$.

The phase shift in total reflection is large, as can be seen from Fig. 3.4; moreover, there is quite a difference between the p- and s-components.

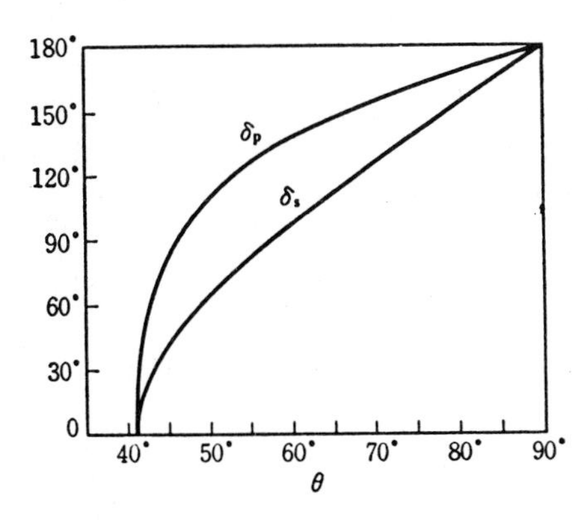

Fig. 3.4. Phase angle of the reflected wave as a function of the incident angle θ for $n_{12} = 1.52$

Calculation from (3.47, 48) shows that the phase difference between the two components is given by

$$\delta_p - \delta_s = 2\tan^{-1}\left(\frac{\cos\theta\sqrt{n_{12}^2\sin^2\theta - 1}}{n_{12}\sin^2\theta}\right). \tag{3.49}$$

For $n_{12} = 1.52$, $\delta_p - \delta_s$ equals 45° ($\pi/4$) when θ is about 47.5° or 55.5°. Thus, by constructing a prism such as shown in Fig. 3.5, so that total reflection takes place on its two faces, the phase difference between the p- and s-components becomes equal to 90°. This prism is called a Fresnel rhomb and it is used as a quarter-wave plate in which the wavelength dependence is small.

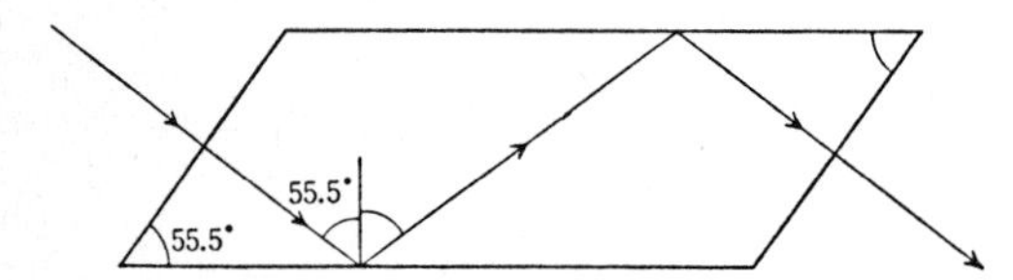

Fig. 3.5. Fresnel rhomb ($n_{12} = 1.52$)

At the boundary between two transparent media, the amplitude reflection coefficient and the transmission coefficient are real quantities, except at total reflection, and their phases are either 0 or π. However, for semiconductors and metals, since $\boldsymbol{j} = \sigma\boldsymbol{E}$ where σ is the electric conductivity, the right-hand side of (3.2) for the frequency ω becomes

$$\boldsymbol{j} + \frac{\partial \boldsymbol{D}}{\partial t} = (\sigma + i\omega\varepsilon')\boldsymbol{E}$$

where $\boldsymbol{D} = \varepsilon'\boldsymbol{E}$. This equation is equivalent to taking the complex permittivity

$$\varepsilon = \varepsilon' - i\varepsilon'' \tag{3.50}$$

with $\varepsilon'' = \sigma/\omega$ and putting $\boldsymbol{j} = 0$. Similarly the index of refraction can be expressed in the complex form

$$\eta = \eta' - i\kappa \tag{3.51}$$

where the real and imaginary parts, when $\mu = \mu_0$, are

$$\eta' = \left\{\frac{1}{2\varepsilon_0}\left[\sqrt{(\varepsilon')^2 + (\varepsilon'')^2} + \varepsilon'\right]\right\}^{1/2}, \tag{3.52}$$

$$\kappa = \left\{\frac{1}{2\varepsilon_0}\left[\sqrt{(\varepsilon')^2 + (\varepsilon'')^2} - \varepsilon'\right]\right\}^{1/2}. \tag{3.53}$$

The quantity κ is called the extinction coefficient.

3.4 Fabry-Perot Resonator

A system of two plane mirrors placed face to face with a certain spacing between them manifests resonance for electromagnetic waves of particular wavelengths. This is the most basic optical resonator and is called the Fabry-Perot resonator.

As shown in Fig. 3.6, the normal to the reflecting surfaces is taken to be the z axis, and the reflecting surfaces are placed at $z = 0$ and $z = L$. The direction of polarization is taken along the x axis and the reflecting surfaces are assumed to be perfect conductors. Generally, an electromagnetic wave propagating in the z direction is given by (3.15). Since $E_x = 0$ at the perfectly conducting surface at $z = 0$, we have $F_1 = -F_2$ in this case. Therefore, if we write $A_x = 2\mathrm{i}F_2$, we have

$$E_x = A_x \mathrm{e}^{\mathrm{i}\omega t} \sin kz \ . \tag{3.54}$$

Since $E_x = 0$ also at $z = L$, we must have $\sin kL = 0$. Consequently only those waves for which

$$k = \frac{n\pi}{L}, \qquad \omega = \frac{n\pi c}{L} \tag{3.55}$$

are permissible. Here n is an integer. When L is very much longer than the wavelength of the light, n is a very large number. Since $k = 2\pi/\lambda$, where λ is the wavelength, (3.55) indicates that L is an integral multiple n of the half-wavelength.

With L kept constant, (3.55) expresses the proper wavenumbers and proper frequencies of the Fabry-Perot resonator. The spatial function of the proper mode is given by $\sin kz$, and the mode number is equal to the number of crests or troughs of the mode function within the interval $0 \leqq z \leqq L$.[3] Any electromagnetic field which varies in space and time can be expressed by a superposition of these modes. Mathematically speaking, an arbitrary function $f(z)$, which becomes 0 at $z = 0$ and $z = L$, can be expanded in terms of

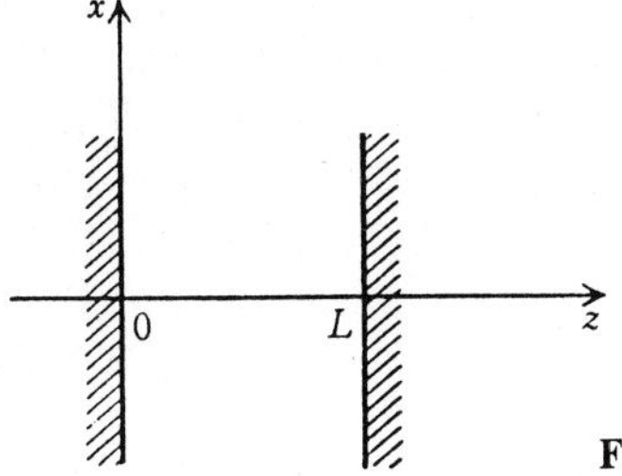

Fig. 3.6. Coordinates in the Fabry-Perot resonator

3 In counting the number of modes the two ends are counted as one.

the mode functions (3.54) having the proper values (3.55).[4] This is the well-known Fourier expansion.

When the surfaces are not perfectly reflecting, E_x is not zero at $z = 0$ and $z = L$. Even in this case, however, it is possible to express the function $f(z)$, representing an arbitrary electromagnetic field within the interval $0 < z < L$, in the form

$$f(z) = \sum_{n=0}^{\infty} \left(A_n \sin \frac{2\pi nz}{L} + B_n \cos \frac{2\pi nz}{L} \right) . \tag{3.56}$$

This equation repeats itself outside the interval $z = 0$–L with periodicity L. For an arbitrary value of z, we have

$$f(z + L) = f(z) .$$

In this case, the mode functions of the electromagnetic field within the interval 0–L are $\sin kz$ and $\cos kz$, where $k = 2n\pi/L$, as can be clearly seen from (3.56). Therefore, comparing (3.56) with (3.55), we see that the mode interval in this case is twice that when the reflection is perfect. However, both are essentially equivalent, since the function has been expanded in terms of the double trigonometric functions, sine and cosine, in this case.

To express the electromagnetic field between imperfectly reflecting surfaces in the form (3.56) is equivalent to determining the modes in accordance with the boundary condition that the electromagnetic fields in the neighborhood of $z = 0$ and $z = L$ are equal to one another. Such a boundary condition is called a periodic boundary condition, and is written for $\boldsymbol{E}(x, y, z, t)$ as

$$\boldsymbol{E}(z = 0) = \boldsymbol{E}(z = L), \quad \frac{\partial \boldsymbol{E}}{\partial z}(z = 0) = \frac{\partial \boldsymbol{E}}{\partial z}(z = L) . \tag{3.57}$$

When we deal with electromagnetic fields in an infinitely wide space, we generally meet with mathematical difficulties such as a diverging integral or an indeterminable product of infinite and infinitesimal quantities. Since, however, we are not concerned with the infinite space of the entire universe in a physical observation, it is convenient to confine our object to the interior of a cube of side L which is larger than the relevant length in an actual observation. In this case we may assume periodic boundary conditions at the faces of the cube. Thus, mathematically, the same electromagnetic field is repeated in the form of a cubic lattice outside this physical space of volume L^3. The treatment of the Fabry-Perot resonator given above is a one-dimensional case. We shall deal with the three-dimensional modes of the electromagnetic field in Chap. 4.

4 If the electromagnetic field at a certain time t is given, the field at any arbitrary time is determined from Maxwell's equations.

3.5 Fabry-Perot Interferometer

The Fabry-Perot interferometer is composed of two plane-parallel reflecting surfaces which are partially transparent. As shown in Fig. 3.7, there are two kinds: (a) two reflecting mirrors held in parallel in a vacuum or filled with gas such as air, and (b) plane-parallel reflecting surfaces of a transparent solid material, such as glass. If we now let monochromatic light enter from different directions and observe the directional distribution of the transmitted light, interference fringes in the form of concentric circles, as shown in Fig. 3.8, are seen. In an actual experiment it is customary to scatter the light from the source by putting a ground-glass plate between the monochromatic light source and the interferometer, and to observe the interference fringes through the interferometer with a low-magnification telescope or to photograph them with a camera focused at infinity.

The reflection coefficients r of the two reflecting surfaces at $z = 0$ and $z = L$ facing each other are assumed to be equal, as are the transmission coefficients d. Although both r and d are functions of the angle of incidence and direction of polarization, they are practically constant[5] when the angle of incidence is small, as can be seen from the Fresnel formula. When the wave vector $\boldsymbol{k}$ of the incident light of frequency ω is in the xz plane, the electric field of light transmitted through the $z = 0$ plane and incident on the $z = L$ plane can, by dropping the term $\exp(\mathrm{i}\omega t)$, be written as

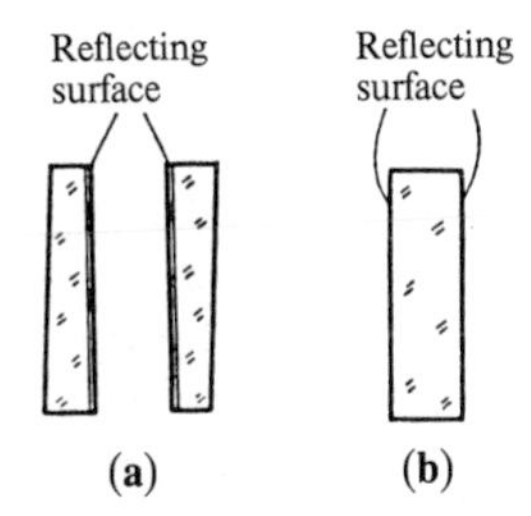

Fig. 3.7 a, b. Fabry-Perot interferometers with (**a**) an air gap, and (**b**) a transparent solid between the reflecting surfaces

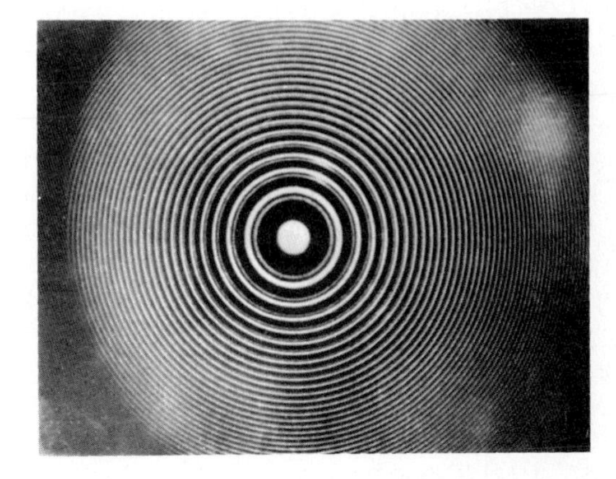

Fig. 3.8. Interference fringes of a Fabry-Perot interferometer. The resolved doubling of the rings shows the presence of two components of slightly different wavelengths in the light source

5 We assume r and d take real values. When there is a shift in phase, it is possible to make r a positive real quantity by considering an equivalent reflecting plane slightly shifted from the geometrical surface.

$$E_i(x, y, z) = E_0 d \exp[-ik(x \sin\theta + z\cos\theta)] , \quad (3.58)$$

where E_0 is the amplitude of the electric field of the light incident on the interferometer.

Since the light is repeatedly reflected at $z = 0$ and $z = L$ and propagates as shown in Fig. 3.9, the electric field of the transmitted light is expressed by the geometric series

$$\begin{aligned} E_d(x, y, z) &= E_0 d^2 \exp[-ik(x\sin\theta + L\cos\theta)] \\ &\times [1 + r^2 \exp(-2ikL\cos\theta) + r^4 \exp(-4ikL\cos\theta) + \ldots] \\ &= \frac{E_0 d^2 \exp[-ik(x\sin\theta + L\cos\theta)]}{1 - r^2 \exp(-2ikL\cos\theta)} , \end{aligned} \quad (3.59)$$

where we have assumed the refractive indices of the media in front and behind the interferometer to be the same. When the refractive index is 1 outside the interferometer and η inside, the angle of incidence outside the interferometer is $\theta_o = \eta\theta$, since θ is the angle of incidence for $0 < z < L$ and $\theta \ll 1$. As the transmitted light intensity is proportional to $|E_d|^2$, we have from (3.59)

$$I_d = \frac{I_0 D^2}{1 + R^2 - 2R\cos(2kL\cos\theta)} , \quad (3.60)$$

where I_0 is the incident light intensity, $R = r^2$ the reflectance, and $D = d^2$ the transmittance.

The transmittance I_d/I_0 of the interferometer varies as shown in Fig. 3.10 when $kL\cos\theta$ changes with frequency ω, separation L between the reflecting planes, or incident angle θ. In this figure the transmittance of the interferometer is given for the case $D = 0.98 - R$, assuming the loss of light at the reflecting plane to be 2%. The transmittance is a maximum for $kL\cos\theta = n\pi$ and a minimum for $(n + 1/2)\pi$, where n is an integer. As can be clearly seen from Fig. 3.10, although the maximum becomes sharper as the reflectance R become large, its height decreases, i.e., the transmittance decreases. If there is no loss whatever at the reflecting planes, we have $D = 1 - R$ and the maximum value of the transmittance is 1 irrespective of the magnitude of R.

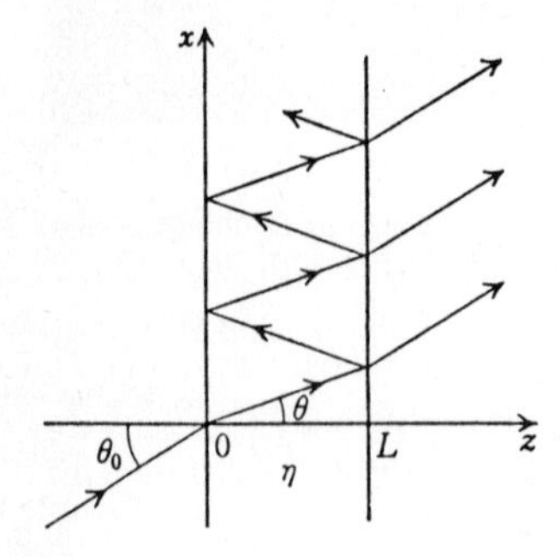

Fig. 3.9. Optical paths in the Fabry-Perot interferometer

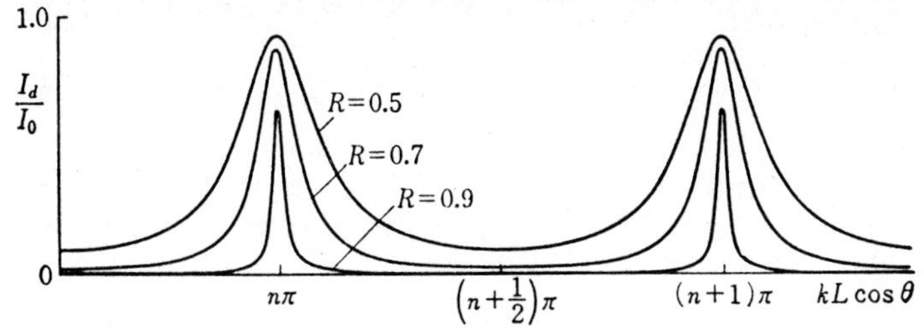

Fig. 3.10. Variation of transmittance as the gap L between the reflecting surfaces or the incident angle θ is changed. Here D is taken equal to $0.98 - R$

In the case of normal incidence $\theta = 0$ and $\cos\theta = 1$. Then the transmittance is a maximum when $kL = n\pi$, and so

$$L = n\frac{\lambda}{2}\,. \tag{3.61}$$

Therefore, as the separation L is changed, maxima of the transmittance can be observed at intervals of half a wavelength. Again if we keep L constant and change ω, the transmittance is a maximum when

$$\omega = n\frac{\pi c}{L}\,.$$

These are equivalent to the resonance conditions (3.55) of the Fabry-Perot resonator.

On examining the variation of the transmittance with angle of incidence θ, keeping L constant, we find that the angle θ_{m} for the maximum transmittance is given by writing

$$kL = n\pi + \phi_0\,, \tag{3.62}$$

where n is an integer, and $0 \leqslant \phi_0 < \pi$, in the form

$$\cos\theta_{\mathrm{m}} = \frac{(n-m)\pi}{n\pi + \phi_0}\,, \quad (m = 0, 1, 2, \ldots)\,.$$

If $L \gg \lambda$, n is very much greater than m and we obtain

$$\theta_{\mathrm{m}} = \sqrt{\frac{2\phi_0}{n\pi} + \frac{2m}{n}}\,, \tag{3.63}$$

using $\cos\theta_{\mathrm{m}} = 1 - \theta_{\mathrm{m}}^2/2$ for $\theta_{\mathrm{m}} \ll 1$. When $\phi_0 = 0$, this becomes

$$\theta_{\mathrm{m}} = \sqrt{\frac{m\lambda}{L}}\,, \tag{3.64}$$

using (3.61). When $\phi_0 \neq 0$, (3.63) may be written as

$$\theta_m = \sqrt{\frac{m\lambda}{L} + \theta_0^2}$$

where θ_0 is the incident angle for $m = 0$. From these results it can be understood that interference fringes of concentric circles, as in Fig. 3.8, are obtained when a monochromatic light source is observed through a Fabry-Perot interferometer.

Since the position of the interference fringes is different for different frequencies, the Fabry-Perot interferometer may be used as a spectrometer. However, we cannot distinguish two frequencies for which the values of kL differ by an integral multiple of π. This is because the interference fringes of different orders of n superpose on one another. An interference fringe of a certain order does not become confused with a fringe of a different order as long as $kL = \omega L/v$ changes only within $\pm\pi/2$. This frequency range is denoted by ω_{FSR} (FSR stands for free spectral range; "free" meaning that it is not hindered by a fringe of another order) and it is expressed as

$$\omega_{FSR} = \frac{\pi v}{L} \ . \tag{3.65}$$

This is the circular frequency corresponding to the spacing of the interference fringes. It is in practice customary to express the FSR in terms of frequency in Hz or wavenumber in cm^{-1}.

The accuracy with which it is possible to measure wavelength or frequency with a Fabry-Perot interferometer is determined by the sharpness of the interference fringes. The finesse $\mathcal{F}$, which is defined below, determines this accuracy. The transmittance is a maximum at $\theta = 0$, and it becomes one half of the maximum value for $\phi_0 = \Delta\phi/2$ if

$$1 + R^2 - 2R\cos\Delta\phi = 2(1-R)^2 \ ,$$

as can be seen from (3.60, 62). Then $\Delta\phi$ is the width of the interference fringe in terms of phase angle. Solving the above equation,we obtain

$$\Delta\phi = \frac{1-R}{\sqrt{R}} \ ,$$

since $\Delta\phi \ll 1$. Thus the width of the interference fringes in terms of frequency is expressed as

$$\Delta\omega = \frac{v}{L}\frac{1-R}{\sqrt{R}} \ . \tag{3.66}$$

We now define the finesse $\mathcal{F}$ by the ratio of the fringe spacing to the fringe width, namely, $\pi/\Delta\phi$ or $\omega_{FSR}/\Delta\omega$, so that

$$\mathcal{F} = \frac{\pi\sqrt{R}}{1-R} \ . \tag{3.67}$$

An ordinary Fabry-Perot interferometer has a finesse of $\mathcal{F} = 10\text{–}100$, whereas a one of high quality may have a finesse greater than 200.

Finally, let us examine the optical electric field in the space between the two reflecting planes. The electric field due to repeated reflection can be written as

$$\begin{aligned} E(x, y, z) &= E_0 d \exp(-ikx \sin\theta) \{\exp(-ikz\cos\theta) \\ &\quad + r\exp[-ik(2L - z)\cos\theta]\} \times (1 + a + a^2 + \dots)\,, \\ a &= r^2 \exp(-2ikL\cos\theta)\,. \end{aligned}$$

The first term in the curly brackets is the wave (3.58), the second term is the wave propagating in the $-z$ direction after reflection at $z = L$, and a is the relative amplitude of the wave which has been reflected twice, at $z = 0$ and L. If we write $b = r\exp(-2ikL\cos\theta)$, we have

$$E(x, y, z) = E_0 d \exp(-ikx\sin\theta) \times \frac{(1 - b)\exp(-ikz\cos\theta) + 2b\cos(kz\cos\theta)}{1 - r^2\exp(-2ikL\cos\theta)}\,. \tag{3.68}$$

Since the denominator is a minimum when

$$kL\cos\theta = n\pi\,, \tag{3.69}$$

where n is an integer, the optical electric field then becomes a maximum. This is the condition for maximum transmittance of the interferometer. The first term in the numerator represents a wave propagating in the z direction and the second term represents a standing wave, where the number of nodes of the standing wave between $z = 0$ and L is given by n.

3.6 Planar Waveguide

In semiconductor lasers and optical integrated circuits, use is made of light propagating along a thin layer of a medium. Among the simplest of such optical waveguides is a thin film whose boundary surfaces are planar and whose refractive index η_1 is greater than the refractive index η_2 of the medium exterior to the film on either side, as shown in Fig. 3.11. In principle this is equivalent to the Fabry-Perot interferometer, but, instead of light being incident almost normal to the surfaces, here it is incident almost parallel to the surfaces. We now take the z axis normal to the surface of the film and consider a wave propagating in the x direction.

If we take the angle of incidence of light coming from the outside (refractive index is 1) towards the film to be θ, as shown in Fig. 3.12, the angle of refraction φ at the end plane of the film is given by $\varphi = \theta/\eta_1$ when $\theta \ll 1$.

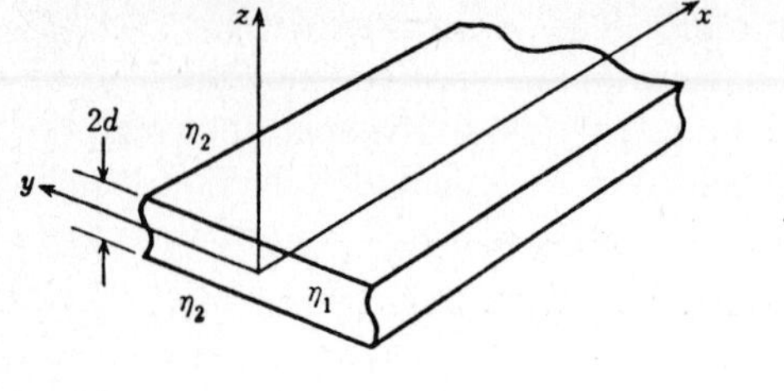

Fig. 3.11. Planar optical waveguide

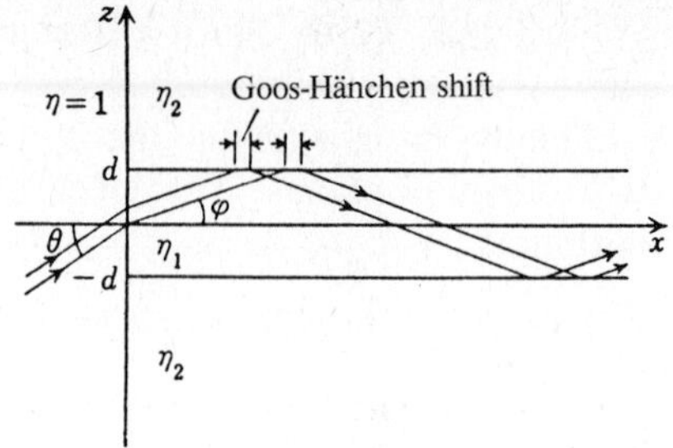

Fig. 3.12. Schematic propagation of light in a planar waveguide

Since the angle of incidence at the boundary plane of the film is $\pi/2 - \varphi$, as shown in Fig. 3.12, the light is totally reflected if

$$\cos\varphi > \frac{\eta_2}{\eta_1} .$$

Approximating $\cos\varphi$ by $1 - \varphi^2/2$ when $\theta \ll 1$ and writing $\Delta\eta = \eta_1 - \eta_2$ for the difference in refractive indices, the light which enters the end plane with an incident angle satisfying the relation

$$\frac{1}{2}\left(\frac{\theta}{\eta_1}\right)^2 < \frac{\Delta\eta}{\eta_1} , \quad \text{i.e.} \quad \theta^2 < 2\eta_1\Delta\eta , \tag{3.70}$$

propagates along the film by repeated total reflection. Its propagation is accompanied by a Goos-Hänchen shift at each total reflection and gives rise to evanescent waves in both sides of the film. Such an optical waveguide is called a slab waveguide or a symmetric planar waveguide [3.2].

The electromagnetic field corresponding to the interference of waves which are totally reflected along the optical waveguide can be determined. By taking the boundary planes of the film to be at $z = d$ and $z = -d$, and with ω being the frequency and k_x the wavenumber of light propagating in the x direction, the optical electric field can be written as

$$\boldsymbol{E}(z)\exp(\mathrm{i}\omega t - \mathrm{i}k_x x) ,$$

since it is uniform in the y direction. From (3.14), its wave equation becomes

$$\frac{d^2\boldsymbol{E}(z)}{dz^2} = (k_x^2 - k^2)\boldsymbol{E}(z) , \tag{3.71}$$

where k is equal to $k_1 = \eta_1\omega/c$ in the interior of the film ($|z| < d$) and $k_2 = \eta_2\omega/c$ outside the film ($|z| > d$). The guided wave along the film thus consists of an evanescent wave outside the film ($|z| > d$) with its amplitude decreasing exponentially with $|z|$, and a standing wave inside the film

($|z| < d$) which can be expressed as a harmonic wave. In order that the solutions of (3.71) should satisfy these conditions, we must have

$$k_x^2 - k_2^2 \equiv \gamma^2 > 0 , \quad k_x^2 - k_1^2 \equiv -\beta^2 < 0 . \tag{3.72}$$

Similar to the microwave waveguide, there are two types of electromagnetic waves propagating along the optical waveguide, namely, the TE wave (transverse electric wave) which has only transverse components for the electric field, i.e., $E_x = 0$, while $H_x \neq 0$ for the magnetic field, and the TM wave (transverse magnetic wave) which has only transverse components for the magnetic field, i.e., $H_x = 0$, while $E_x \neq 0$ for the electric field.

We shall deal first with the TM wave. We may write

$$E_x = A \sin \beta z \quad \text{or} \quad B \cos \beta z \tag{3.73}$$

inside the film ($|z| < d$), and

$$E_x = C\,\mathrm{e}^{-\gamma|z|} \tag{3.74}$$

outside ($|z| > d$), where we have dropped the factor $\exp(\mathrm{i}\omega t - \mathrm{i}k_x x)$. Since A, B, and C are constants and E_x is continuous on the boundary plane, we have

$$C = A\,\mathrm{e}^{\gamma d} \sin \beta d \quad \text{or} \quad B\,\mathrm{e}^{\gamma d} \cos \beta d . \tag{3.75}$$

Again, since the film is assumed to extend in the y direction as well, E_y and H_x are both 0, and hence the other components E_z and H_y can be expressed in terms of E_x as shown below.

Inside the film, the y component of (3.1) and z component of (3.2) of Maxwell's equations are, respectively,

$$\frac{dE_x}{dz} + \mathrm{i}k_x E_z = -\mathrm{i}\omega\mu_0 H_y , \tag{3.76}$$

$$-\mathrm{i}k_x H_y = \mathrm{i}\omega\varepsilon_1 E_z , \tag{3.77}$$

with $\varepsilon_1 = \eta_1^2 \varepsilon_0$. Eliminating H_y from these two equations, we have

$$\frac{dE_x}{dz} + \mathrm{i}k_x E_z = \mathrm{i}\frac{k_1^2}{k_x} E_z .$$

Then, using $\beta^2 = k_1^2 - k_x^2$, we obtain

$$E_z = -\mathrm{i}\frac{k_x}{\beta^2}\frac{dE_x}{dz} . \tag{3.78}$$

Also from (3.77) we obtain

$$H_y = \mathrm{i}\frac{\omega\varepsilon_1}{\beta^2}\frac{dE_x}{dz} . \tag{3.79}$$

Outside the film, ε_1 is replaced by ε_2 and we obtain, using $\gamma^2 = k_x^2 - k_2^2$,

$$E_z = \mathrm{i}\frac{k_x}{\gamma^2}\frac{dE_x}{dz}, \tag{3.80}$$

$$H_y = -\mathrm{i}\frac{\omega\varepsilon_2}{\gamma^2}\frac{dE_x}{dz}. \tag{3.81}$$

The electromagnetic fields thus obtained can be written (i) inside the film ($|z| < d$), from (3.73, 78, 79), as

$$\begin{cases} E_x = A\sin\beta z \\ E_z = -\mathrm{i}A\dfrac{k_x}{\beta}\cos\beta z \\ H_y = \mathrm{i}A\dfrac{\omega\varepsilon_1}{\beta}\cos\beta z \end{cases} \quad \text{or} \quad \begin{cases} E_x = B\cos\beta z \\ E_z = \mathrm{i}B\dfrac{k_x}{\beta}\sin\beta z \\ H_y = -\mathrm{i}B\dfrac{\omega\varepsilon_1}{\beta}\sin\beta z \end{cases}$$

and (ii) outside the film ($z > d$), from (3.74, 80, 81), as

$$\begin{cases} E_x = A\,\mathrm{e}^{-\gamma(z-d)}\sin\beta d \\ E_z = -\mathrm{i}A\dfrac{k_x}{\gamma}\mathrm{e}^{-\gamma(z-d)}\sin\beta d \\ H_y = \mathrm{i}A\dfrac{\omega\varepsilon_2}{\gamma}\mathrm{e}^{-\gamma(z-d)}\sin\beta d \end{cases} \quad \text{or} \quad \begin{cases} E_x = B\,\mathrm{e}^{-\gamma(z-d)}\cos\beta d \\ E_z = -\mathrm{i}B\dfrac{k_x}{\gamma}\mathrm{e}^{-\gamma(z-d)}\cos\beta d \\ H_y = \mathrm{i}B\dfrac{\omega\varepsilon_2}{\gamma}\mathrm{e}^{-\gamma(z-d)}\cos\beta d\,. \end{cases}$$

These equations can also be used for points outside the film where $z < -d$, by writing $|z|$ for z. For a given film the parameters β, γ, and k are determined from the frequency and the mode of the guided wave as explained below.

Since H_y is continuous at the surface of the film ($z = d$), the following condition must be satisfied:

$$\frac{\varepsilon_1}{\beta}\cos\beta d = \frac{\varepsilon_2}{\gamma}\sin\beta d \quad \text{or} \quad -\frac{\varepsilon_1}{\beta}\sin\beta d = \frac{\varepsilon_2}{\gamma}\cos\beta d\,.$$

Hence,

$$\tan\beta d = \frac{\varepsilon_1}{\varepsilon_2}\frac{\gamma}{\beta} \quad \text{or} \quad -\frac{\varepsilon_2}{\varepsilon_1}\frac{\beta}{\gamma}. \tag{3.82}$$

From (3.72) the maximum value β_m of β is given by

$$\beta_\mathrm{m}^2 = k_1^2 - k_2^2$$

and we find $\gamma^2 = \beta_m^2 - \beta^2$. Then the condition (3.82), that is, the characteristic equation, becomes

$$\frac{\varepsilon_1}{\varepsilon_2}\frac{\sqrt{\beta_m^2-\beta^2}}{\beta} = \tan\beta d \quad \text{or} \quad -\cot\beta d\,. \tag{3.83}$$

A diagrammatic method of solving (3.83) is given in Fig. 3.13. The left-hand side of (3.83) is a function independent of the thickness of the film and is shown as a thick line. The functions $\tan\beta d$ and $-\cot\beta d$ are periodic functions with periodicity π/d and are, respectively, shown as full and dotted lines. Fig. 3.13 shows an example when $d = 3.2/\beta_m$.

For the sine-function mode $E_x = A\sin\beta z$, the solutions of (3.83) are given by intersections of the thick line with the full lines. We find two such intersections in Fig. 3.13. The solutions, starting with the lowest value, are denoted by $\beta_0, \beta_2, \ldots$, where β_0 is a little less than $\pi/2$ and β_2 is a little less than $3\pi/2$. Thus the distribution of the transverse components of the electromagnetic fields, H_y and E_z, is given by the TM_0 wave in Fig. 3.14 for $\beta = \beta_0$, and the TM_2 wave for $\beta = \beta_2$. As the film becomes thicker, there will be more solutions, since the period of $\tan\beta d$ becomes shorter with respect to β.

The solutions for the cosine-function mode $E_x = B\cos\beta z$ are given by intersections of the thick line with the dotted lines in Fig. 3.13. The transverse components H_y and E_z of the electromagnetic field of this mode are odd functions of z and the solutions are denoted by $\beta_1, \beta_3, \ldots$, of which the TM_1 wave for $\beta = \beta_1$ is shown in Fig. 3.14. When the film is very thin there

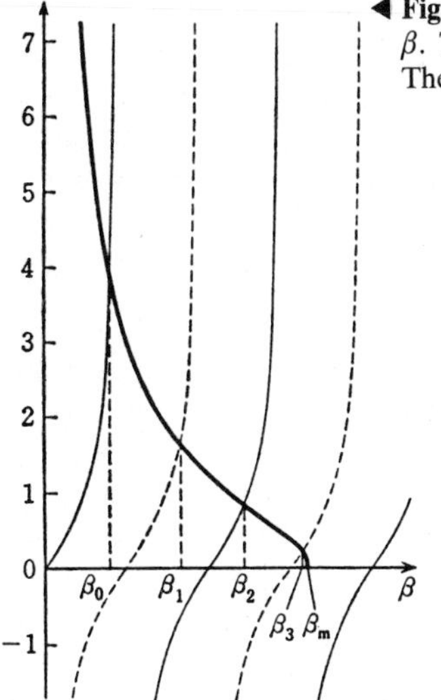

◀ **Fig. 3.13.** The thick line shows the left-hand side of (3.83) vs. β. Thin lines give $\tan\beta d$ and the dashed lines give $-\cot\beta d$. Therefore, $\beta_0, \beta_1, \ldots$ are solutions of (3.83)

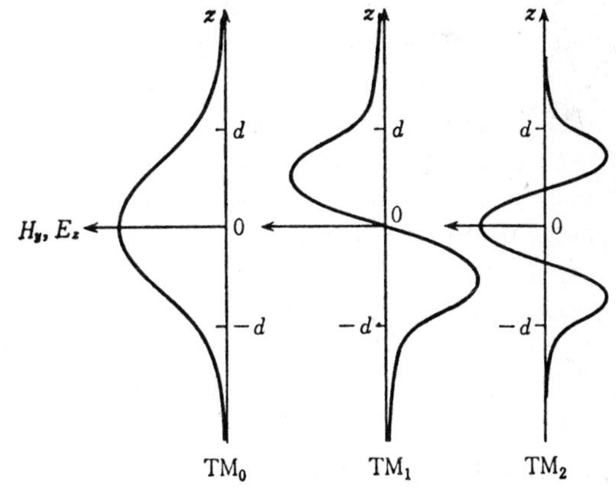

Fig. 3.14. Amplitude distribution for each mode of a thin-film optical waveguide

will be only the TM_0 wave as can be seen from Fig. 3.13. For a film of given thickness $2d$ the TM_0 wave propagates, however low the frequency of the light may be, whereas the TM_n waves with $n = 1, 2, \ldots$ can only propagate when the frequency is higher than their respective cut-off frequencies.

The above equations for the TM waves will hold for the TE waves if $\boldsymbol{E}$ and $\boldsymbol{H}$ are interchanged and the constants duly modified. The characteristic equation determining the propagation constants (β, γ, and k_x) of the even or odd modes of TE waves is given by

$$\frac{\sqrt{\beta_{\mathrm{m}}^2 - \beta^2}}{\beta} = \tan \beta d \quad \text{or} \quad -\cot \beta d \tag{3.84}$$

instead of (3.83), while the conditions (3.72) remain unchanged. The coefficient $\varepsilon_1/\varepsilon_2$ in (3.83) is equal to $(\eta_1/\eta_2)^2 = n_{12}^2$ and is hence greater than 1, but it is not very different from 1. Therefore the z distribution of E_y for the TE_n wave is almost the same as the distribution of H_y for the TM_n wave shown in Fig. 3.14.

There is no cut-off frequency for either the TE_0 or TM_0 wave, but the longer the wavelength compared to d the smaller the value of γ, resulting in the larger portion of the light energy propagating outside the film. Denoting the cut-off frequency of the TE_n and TM_n waves by ν_n, we obtain, by putting $k_x = 0$ and $\beta = \beta_n$ in (3.83, 84),

$$\tan \beta_n d = 0 \quad \text{or} \quad \cot \beta_n d = 0 \ ,$$

so that

$$\nu_n = n \frac{v_1}{4d} \ .$$

In terms of the wavelength in the medium with refractive index $\eta_1 = c/v_1$, the cut-off wavelength becomes

$$\lambda_n = \frac{4d}{n}$$

namely, the thickness $2d$ of the film is equal to an integral multiple of one half the cut-off wavelength.

In most optical waveguides in practice, the refractive index of the medium above the film is different from that below the film. Furthermore, in order to restrict the distribution in the transverse direction, a waveguide with a cross-section as shown in Fig. 3.15 is used. It is not easy to describe the electromagnetic field of light propagating in a waveguide of rectangular cross-section, but the basic characteristics can be understood by applying the results described above. The same may be said of optical fibers. In optical fibers with a circular cross-section it is convenient to use cylindrical coordinates (r, θ, z) and treat the wave equation where the refractive index $\eta(r)$ is a function of the radius. In the case where the distribution of refractive index is

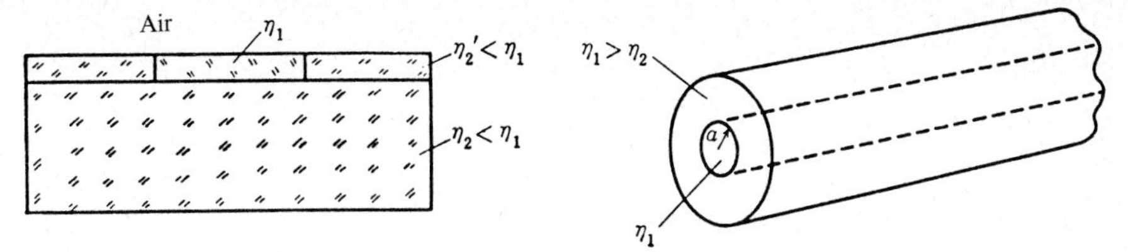

Fig. 3.15. Optical waveguide of rectangular cross-section

Fig. 3.16. Optical fiber

a step function such as $\eta(r) = \eta_1$ for $r < a$ and $\eta(r) = \eta_2$ ($< \eta_1$) for $r > a$ (Fig. 3.16), the solution of the wave equation is expressed in terms of Bessel and Neumann functions [3.3].

3.7 Gaussian Beam

In the treatment of the wave equation (3.8) in this chapter we have so far considered the electric and magnetic fields as vector quantities satisfying Maxwell's equations. However, in the ordinary treatment of wave optics it is sufficient to use the wave equation of a scalar variable u,

$$\nabla^2 u + k^2 u = 0 , \tag{3.85}$$

for the explanation of diffraction, interference, birefringence, etc. This is equivalent to taking only one of the components of the vector. Equation (3.85) is known as the Helmholtz equation. In general, when the extension of the medium is large compared to the wavelength, the light wave is almost purely transverse so that a scalar treatment is quite satisfactory. In this section we shall use the Helmholtz equation to analyze the characteristics of monochromatic beam of light. It is well known that any arbitrary electromagnetic field can be expanded into plane or spherical waves, but a light beam propagating along an arbitrary axis can be approximately expanded into modes of Hermite-Gaussian functions along that axis, as shown below.

Taking the z axis along the light beam and the wavenumber of the medium for transverse waves as k, we put

$$u = f(x, y, z)\, e^{-ikz} . \tag{3.86}$$

The function f of a solution representing the light beam must become practically zero for large values of x or y and change only gradually with z. By neglecting the second-order derivative of f with respect to z, the Helmholtz equation (3.85) becomes

$$\frac{\partial^2 f}{\partial x^2} + \frac{\partial^2 f}{\partial y^2} - 2ik\frac{\partial f}{\partial z} = 0 . \tag{3.87}$$

To solve (3.87) by the method of separation of variables, we shall try putting

$$f(x, y, z) = X(x)\,Y(y)\,G(z)\exp\left(-\frac{x^2+y^2}{F(z)}\right). \tag{3.88}$$

Here $G(z)$ and $F(z)$ are slowly varying functions and the last term is one which becomes practically zero at far off-axis points where $x^2 + y^2 = r^2$ is large. The justification of our assumption is shown below by the fact that the solution of (3.87) can be obtained with this trial solution (3.88).

We differentiate (3.88) to obtain

$$\frac{\partial^2 f}{\partial x^2} = \left(X'' - \frac{4x}{F}X' - \frac{2}{F}X + \frac{4x^2}{F^2}X\right) YG\,\exp\left(-\frac{x^2+y^2}{F}\right),$$

$$\frac{\partial f}{\partial z} = \left[G' + \frac{GF'}{F^2}(x^2+y^2)\right] XY\exp\left(-\frac{x^2+y^2}{F}\right),$$

where $dX(x)/dx$, etc., have been abbreviated to X', etc. Since the y derivatives take the same form as the x derivatives, by substituting them into (3.87) we obtain

$$\frac{X''}{X} - \frac{4x}{F}\frac{X'}{X} + \frac{Y''}{Y} - \frac{4y}{F}\frac{Y'}{Y} - \frac{4}{F} - 2ik\frac{G'}{G} + 2\,\frac{x^2+y^2}{F^2}(2 - ikF') = 0\,. \tag{3.89}$$

In order that this equation should hold with X as a function of x only, Y as a function of y only, and F and G as functions of z only, it is necessary first of all that $2 - ikF' = 0$. We integrate this to obtain

$$F(z) = \frac{2}{ik}(z + C)\,. \tag{3.90}$$

The integration constant C is, in general, a complex quantity. If the wave front is a plane at $z = z_0$ (the phase of f is independent of x and y), we may write

$$\frac{2}{ik}C = w_0^2 - \frac{2}{ik}z_0\,. \tag{3.91}$$

The radius at which the last term in (3.88) is equal to 1/e becomes a minimum w_0 at $z = z_0$. Hence w_0 represents the beam waist and is called the minimum beam radius.

Once $F(z)$ is given by (3.90), we can separate (3.89) into ordinary differential equations in $X(x)$, $Y(y)$, and $G(z)$. The Hermitian differential equation for X is

$$\frac{X''}{X} - \frac{4x}{F}\frac{X'}{X} + \frac{4n}{F} = 0$$

with a similar equation for Y. The solutions are, respectively,

$$X(x) = \mathrm{H}_n\left(\sqrt{\frac{2}{F}}x\right), \quad Y(y) = \mathrm{H}_m\left(\sqrt{\frac{2}{F}}y\right) \tag{3.92}$$

where H denotes the Hermitian polynomial of integral order n or m. Using (3.90, 92) we can obtain the following differential equation for G from (3.89):

$$\frac{G'}{G} = \frac{2\mathrm{i}}{k} \times \frac{n+m+1}{F} = -\frac{n+m+1}{z+C}. \tag{3.93}$$

This is integrated to obtain

$$G(z) = \frac{A}{(z+C)^{n+m+1}} \tag{3.94}$$

where A is an integration constant. We have now obtained the solution of the Helmholtz equation in the form of a product of Hermitian polynomials and the Gaussian distribution function, as

$$u(x,y,z) = \mathrm{H}_n\left(\sqrt{\frac{2}{F}}x\right)\mathrm{H}_m\left(\sqrt{\frac{2}{F}}y\right) \times \frac{A}{(z+C)^{n+m+1}} \exp\left(-\frac{x^2+y^2}{F} - \mathrm{i}kz\right), \tag{3.95}$$

where F and C are given by (3.90, 91).

If cylindrical coordinates are used in solving the Helmholtz equation for a light beam which is distributed only in the neighborhood of an axis, the proper modes can be expressed in the form of a product of Laguerre polynomials and a Gaussian distribution function. In any case the lowest-order mode is a Gaussian distribution. The term "Gaussian beam" usually means a light beam of this lowest-order mode.

Let us now collect together the basic characteristics of the Gaussian beam which are important in laser optics. The solution for the fundamental mode is given by putting $n = m = 0$ in the above equation. In this case (3.95) gives a Gaussian distribution for x and y. Since $F(z)$ is a complex quantity, by separating $1/F(z)$ into real and imaginary parts, we obtain from (3.90, 91)

$$\frac{1}{F(z)} = \frac{k^2 w_0^2 + 2\mathrm{i}k(z-z_0)}{k^2 w_0^4 + 4(z-z_0)^2}. \tag{3.96}$$

Then, the beam radius as a function of z is

$$w(z) = w_0 \sqrt{1 + \frac{4(z - z_0)^2}{k^2 w_0^4}} \tag{3.97}$$

and the phase factor in (3.96) can be rewritten, together with $\exp(-ikz)$ in (3.96), as

$$\exp\left[-ik\left(z + \frac{2(z - z_0)\, r^2}{k^2 w_0^4 + 4(z - z_0)^2}\right)\right]$$

where $x^2 + y^2 = r^2$. This represents the curvature of the wave surface, as shown in Fig. 3.17. Since the displacement from the plane at z is expressed by

$$\frac{2(z - z_0)\, r^2}{k^2 w_0^4 + 4(z - z_0)^2} = \frac{r^2}{2R},$$

the radius of curvature R of the wave surface is obtained to be

$$R(z) = (z - z_0)\left(1 + \frac{k^2 w_0^4}{4(z - z_0)^2}\right). \tag{3.98}$$

R is infinite at $z = z_0$ and becomes $R \approx z - z_0$ as z recedes far from z_0.

The complex amplitude of the Gaussian beam on the axis is expressed by $G(z)$ which, from (3.90, 94), becomes

$$G(z) = \frac{2A}{ikF(z)}$$

for $n = m = 0$. From (3.96), the phase angle $\phi(z)$ is given by

$$\tan\phi = -\frac{k w_0^2}{2(z - z_0)}. \tag{3.99}$$

Then, by changing the constant to $A_0 = 2A/kw_0$, we have

$$G(z) = \frac{A_0}{w} e^{i\phi}.$$

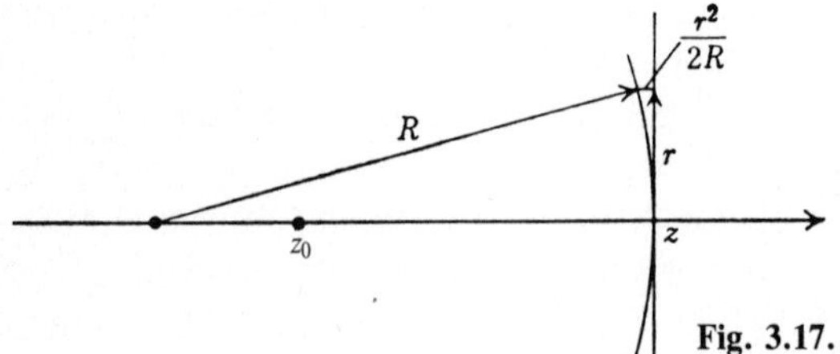

Fig. 3.17. Deviation of the wave surface and the radius of curvature R

Thus, the amplitude of the Gaussian beam at $r = \sqrt{x^2 + y^2}$ can finally be written as

$$u(r, z) = \frac{A_0}{w} \exp\left[-\frac{r^2}{w^2} - ik\left(z + \frac{r^2}{2R}\right) + i\phi\right] \tag{3.100}$$

where w, R, and ϕ are functions which change slowly with z as calculated above.

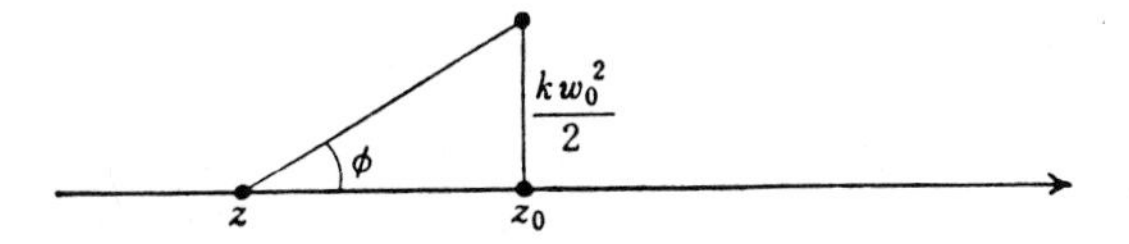

Fig. 3.18. Relation between the phase angle ϕ and the position z

Figure 3.18 shows the relation between the phase angle and z. From (3.97) it follows that the length of the sloping line is proportional to w. As z is increased from $-\infty$, ϕ increases from 0, becomes 90° at $z = z_0$, and 180° at $z = +\infty$. As noted above, the beam radius is smallest at $z = z_0$, where the wave surface (surface of equal phase) becomes a plane. However, the wavelength in this neighborhood is longer than that of a plane wave which is extended in the transverse directions. The wave surfaces of the Gaussian beam and the light rays normal to the wave surfaces are shown in Fig. 3.19.

At a point z, distant from the beam waist at z_0, the beam radius is proportional to $z - z_0$, and the half angle $\Delta\theta$ where the amplitude of the Gaussian beam becomes $1/e$ is obtained from (3.97) to be

$$\Delta\theta = \frac{2}{kw_0} = \frac{\lambda}{\pi w_0}$$

at large distances. Thus, the product of the minimum beam diameter $2w_0$ and the beam spread $2\Delta\theta$ at distance far from z_0 is $(4/\pi)\lambda$, i.e., the product is equal to about one wavelength.

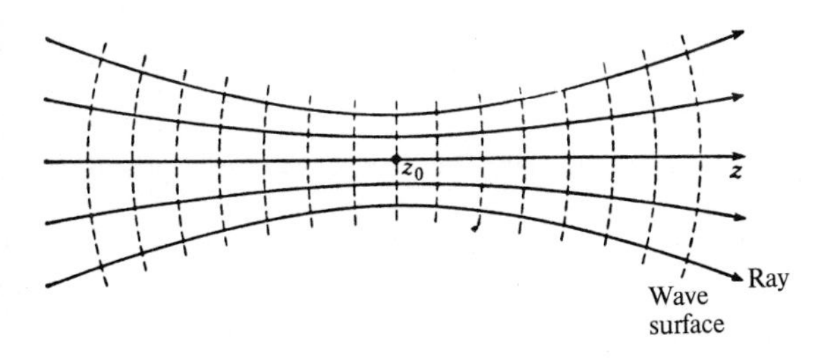

Fig. 3.19. Light rays (—) and the wave surfaces (---) in a Gaussian beam

Problems

3.1 Find the reflection coefficient at normal incidence on a plane boundary between two media having refractive indices η_1 and η_2, respectively. Why has r_p in (3.28) the opposite sign to r_s in (3.33) at normal incidence?

Answer: $r_s = -r_p = (\eta_1 - \eta_2)/(\eta_1 + \eta_2)$. See Fig. 3.1.

3.2 Verify from Maxwell's equations that the electric and magnetic fields of a plane electromagnetic wave in a homogeneous isotropic medium are, in general, perpendicular to each other irrespective of the dielectric and magnetic losses.

Hint: Show that $\boldsymbol{E} \cdot \boldsymbol{H} = 0$.

3.3 Consider a monochromatic plane electromagnetic wave in a medium with a complex index of refraction (3.51). Find a formula which gives the phase difference between the magnetic and the electric fields.

Answer: $\varphi_H - \varphi_E = -\tan^{-1}(\kappa/\eta')$.

3.4 Find an expression for the reflection coefficient of the p-component, to first order, for a small deviation $\Delta\theta$ from the Brewster angle. Evaluate the reflectance for $\Delta\theta = 1°$ and $n_{21} = 1.52$.

Answer: $r_p = -\dfrac{n_{21}^4 - 1}{2n_{21}^3}\Delta\theta$ and $R_p = 1.16 \times 10^{-4}$.

3.5 Find an expression for the reflection coefficient, to first order for a small angle of tilt of the plane of polarization $\Delta\varphi$, when the angle of incidence is exactly equal to the Brewster angle. Evaluate the reflectance for $\Delta\varphi = 1°$ and $n_{21} = 1.52$.

Answer: $(\Delta\varphi r_s)^2 = 4.8 \times 10^{-5}$.

3.6 Show that the resolving power of a Fabry-Perot interferometer used as a spectrometer is given by

$$\frac{\lambda}{\Delta\lambda} = n\mathcal{F}$$

where n is the integer in (3.61), $\mathcal{F}$ the finesse, and $\Delta\lambda$ the difference in wavelength corresponding to $\Delta\phi$. Rewrite the expression in terms of L, λ, and R and calculate the resolving power for $\lambda = 0.57$ μm, $L = 6$ cm, and $R = 0.91$.

Answer: 7.0×10^6.

3.7 Show that the reflectance at normal incidence on a glass surface coated with a thin film will be zero if the film has an optical thickness of one quarter-wavelength and the refractive index of the film equals the geometric mean of the refractive indices of air and glass. Draw a graph showing the wavelength dependence of the reflectance of the glass surface coated for antireflection at a wavelength in air of 500 nm. Show qualitatively how the graph should be modified when normal dispersions of the glass and the film are taken into account.

Remark: $\partial\eta/\partial\lambda < 0$ in normal dispersion.

3.8 Find a formula that gives the maximum angle of acceptance θ_{max} without assuming $\theta^2 \ll 1$, and evaluate the angle θ_{max} in degrees when $\eta_1 = 1.48$ and $\eta_2 = 1.44$.

Answer: $\sin\theta_{max} = \sqrt{\eta_1^2 - \eta_2^2}$, $\theta_{max} = 20°$.

3.9 Explain graphically why the evanescent component of light traveling along a planar waveguide attenuates more slowly for higher modes.

Hint: Refer to Fig. 3.13 and (3.72).

3.10 Draw a graph showing the distribution of light intensity along the axis of a Gaussian laser beam.

Hint: $I(z) = [w_0/w(z)]^2 I_0$.

3.11 Assuming that the probability of a two-photon transition is proportional to the square of the light intensity or the fourth power of the optical field, show that the total number of two-photon transitions induced by a focused beam of Gaussian light in a large material is essentially independent of the beam radius at the focus. Explain why this number actually decreases when the focal length of the lens is too short or too long.

3.12 Draw a graph showing the radius of curvature of the wave surface in a Gaussian beam as a function of z, and find the minimum radius of curvature.

Answer: A hyperbola having asymptotes $z - z_0 = 0$ and $z - z_0 = R$; $|R|_{min} = kw_0^2$ at $z = z_0 \pm kw_0^2/2$.

3.13 Derive a formula that gives the transmittance of a Gaussian beam through a cylindrical tube of length $2L$ and radius a, assuming that the beam waist is at the center of the tube and that the power incident on the entrance cross-section πa^2 can be transmitted. Then show that the maximum transmittance is obtained when the beam waist is equal to $\sqrt{\lambda L/2\pi}$. What percentage of the Gaussian beam of light at 0.63 μm and 10.6 μm can respectively transmit through a tube of $a = 1$ mm and $L = 10$ cm?

Answer: 99%, 16.5%.

4. Emission and Absorption of Light

It was from the study of the emission and absorption of light that such subjects as spectrochemical analysis, atomic physics and quantum mechanics grew and led further to the inventions of the maser and the laser. It was eventually found that the emission and absorption of light by matter took place discretely with an energy quantum or a photon as a unit, and it was this fact which became the foundation of the development of quantum physics during this century. Before deriving Planck's law of thermal radiation which first showed the necessity of using the quantized energy of light, we shall calculate the mode density of electromagnetic waves.

4.1 Mode Density of Electromagnetic Waves

In the case of an electromagnetic field in a closed space or in a laser resonator in which light is effectively confined, it is most convenient to expand the electromagnetic field in terms of the characteristic modes of the closed space or the laser resonator. On the other hand, when we treat the emission of light from a body into free space, we must consider a continuous distribution of modes in a space of infinite extension. Even in this case, however, by limiting the electromagnetic field to a sufficiently large finite space, as stated in Sect. 3.4, the treatment is simplified, since the characteristic modes then become discrete. When using Cartesian coordinates, it is customary to take a cube of side L in each of the x, y, and z directions. It is obvious that L must be much greater than both the size of the body and the wavelength of the light, but when considering the interaction between an atom and light, it is also necessary that $L \gg c\tau_a$, where τ_a is the lifetime of the excited state of the atom.

Let us now consider the characteristic modes of the electromagnetic waves in a cube of such dimensions and calculate the number of modes per unit volume of space. Let the unit vectors in the Cartesian coordinates be $\hat{\boldsymbol{x}}$, $\hat{\boldsymbol{y}}$, and $\hat{\boldsymbol{z}}$; then the wave vector $\boldsymbol{k}$ of a plane wave in an arbitrary direction is expressed as

$$\boldsymbol{k} = k_x \hat{\boldsymbol{x}} + k_y \hat{\boldsymbol{y}} + k_z \hat{\boldsymbol{z}} \,. \tag{4.1}$$

The complex amplitude of the plane electromagnetic wave having this wave vector can be written as

$$A \exp(\mathrm{i}\omega t - \mathrm{i}\boldsymbol{k} \cdot \boldsymbol{r}) ,$$

where A is a complex quantity expressing the magnitude and phase of the amplitude, ω is the frequency, and $\boldsymbol{r}$ is the radius vector $\boldsymbol{r} = x\hat{\boldsymbol{x}} + y\hat{\boldsymbol{y}} + z\hat{\boldsymbol{z}}$.

Let us first calculate the density of modes satisfying the periodic boundary conditions on the surface of a cube of side L. Since the boundary conditions require the amplitude of the wave to be the same at $x = 0$ and $x = L$, the characteristic mode is determined from

$$\exp(\mathrm{i}k_xL) = 1 , \quad \text{therefore} \quad k_x = \frac{2\pi}{L}n_x .$$

Similarly, we have

$$k_y = \frac{2\pi}{L}n_y , \quad k_z = \frac{2\pi}{L}n_z ,$$

where n_x, n_y, and n_z are integers which may be positive, negative, or zero. We calculate $k^2 = k_x^2 + k_y^2 + k_z^2$ to be

$$k^2 = \left(\frac{2\pi}{L}\right)^2 (n_x^2 + n_y^2 + n_z^2) . \tag{4.2}$$

If we assume the space to be a vacuum for the sake of simplicity, the frequency $\omega = kc$ of light designated by n_x, n_y, and n_z is given by

$$\omega^2 = \left(\frac{2\pi c}{L}\right)^2 (n_x^2 + n_y^2 + n_z^2) . \tag{4.3}$$

Since n_x, n_y, and n_z are discrete, ω also takes discrete values. However, as L is very much greater than the wavelength of light ($\lambda = 2\pi/k$), $n_x^2 + n_y^2 + n_z^2$ is a very large quantity of the order of L^2/λ^2. Since there are a very large number of densely populated modes, we can calculate the number of modes in $\boldsymbol{k}$ space in the following manner.

Because k_x, k_y, and k_z are all integral multiples of $2\pi/L$, as stated above, by taking n_x, n_y, and n_z, respectively, in the x, y, and z directions as in Fig. 4.1, the number of sets of n_x, n_y, and n_z which lie within a sphere of radius R is equal to the volume $(4\pi/3)R^3$ of the sphere. Since there are two modes of electromagnetic waves of different directions of polarization to each set (n_x, n_y, n_z), the total number of modes represented by points in the sphere of radius R is equal to $2 \times (4\pi/3)R^3$. By using $k = (2\pi/L)R$ and $\omega = (2\pi c/L)R$ for $\sqrt{n_x^2 + n_y^2 + n_z^2} = R$, the total number of characteristic modes of the electromagnetic waves whose frequency lies within the range from 0 to ω is given by

$$2 \times \frac{4\pi}{3}\left(\frac{\omega L}{2\pi c}\right)^3 = \frac{\omega^3}{3\pi^2c^3}L^3 . \tag{4.4}$$

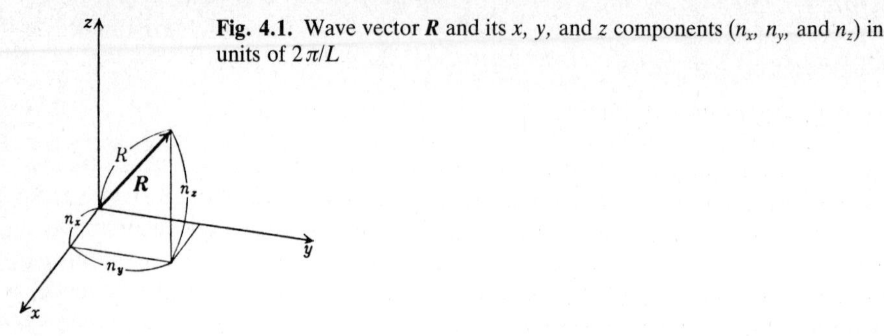

Fig. 4.1. Wave vector $\boldsymbol{R}$ and its x, y, and z components (n_x, n_y, and n_z) in units of $2\pi/L$

By differentiating this, we find that the number of modes with frequencies between ω and $\omega + d\omega$ is given by

$$\frac{\omega^2}{\pi^2 c^3} L^3 d\omega \ .$$

Since the volume of the cube is L^3, the number of modes per unit volume of space, namely, the mode density within the frequency range between ω and $\omega + d\omega$ is expressed as

$$m(\omega)\, d\omega = \frac{\omega^2}{\pi^2 c^3} d\omega \ . \tag{4.5}$$

It should be noted that not only is this true for periodic boundary conditions, but it also holds when the boundaries of the cube are perfect conductors. The modes of the electromagnetic waves in this case can be regarded as a three-dimensional extension of the Fabry-Perot resonator modes, described in Sect. 3.4, so that the characteristic modes can be expressed as a combination of standing waves in each of the x, y, and z directions. Since the standing wave in a certain direction is a superposition of two waves propagating respectively in the $+\boldsymbol{k}$ direction and in the $-\boldsymbol{k}$ direction, a change in the sign of $\boldsymbol{k}$ leads to an identical standing wave. Thus, in counting the number of standing-wave modes, we take only positive integers for n_x, n_y, and n_z. From the boundary conditions at $x = 0$ and L, $y = 0$ and L, and $z = 0$ and L, we must have

$$k_x = \frac{\pi}{L} n_x \ , \quad k_y = \frac{\pi}{L} n_y \ , \quad k_z = \frac{\pi}{L} n_z$$

just as in (3.55) in Sect. 3.4. We therefore obtain

$$\omega^2 = \left(\frac{\pi c}{L}\right)^2 (n_x^2 + n_y^2 + n_z^2) \ .$$

Thus, the number of standing-wave modes within the frequency range from 0 to ω is obtained by calculating one eighth of the volume of the sphere of radius $R = \omega L/\pi c$, where n_x, n_y, and n_z are all positive integers, and multiplying the result by 2 for the two directions of polarization. The result is

$$\frac{2}{8} \times \frac{4\pi}{3}\left(\frac{\omega L}{\pi c}\right)^3 = \frac{\omega^3}{3\pi^2 c^3} L^3 ,$$

which is the same as (4.4). Consequently, the mode density of electromagnetic waves in a space surrounded by perfect conductors is given by (4.5) as before.

If, instead of using the circular frequency ω, we calculate the mode density in terms of linear frequency $\nu = \omega/2\pi$ in the range between ν and $\nu + d\nu$, (4.5) is rewritten as

$$m(\nu)\, d\nu = \frac{8\pi\nu^2}{c^3} d\nu . \tag{4.6}$$

4.2 Planck's Law of Thermal Radiation

According to statistical mechanics, the probability that the energy of an oscillator in thermal equilibrium lies between U and $U + dU$ is given by

$$p(U)\, dU = \frac{1}{k_B T} \exp(-U/k_B T) . \tag{4.7}$$

This is known as a canonical distribution, and $k_B = 1.38 \times 10^{-23}$ J/K is the Boltzmann constant. At a given temperature, the probability distribution decreases with increasing energy, as shown in Fig. 4.2. At low temperatures the probability distribution is concentrated mainly at low energies. At high temperatures, on the other hand, it is more evenly spread out so that it has an appreciable distribution at high energies.

The so-called black-body radiation can be regarded as electromagnetic waves of light in thermal equilibrium with a body at a high temperature. It then follows that the energy of electromagnetic waves in each mode takes a canonical distribution (4.7), and so long as the energy U of the electromagnetic waves with frequencies of the characteristic modes is considered to take continuous values, it is not in any way possible to explain the observed spectral distribution of the black-body radiation.

In 1900 *M. Planck* assumed that the energy U of electromagnetic waves, instead of taking continuous values, takes only the discrete values

$$U = nh\nu = n\hbar\omega , \tag{4.8}$$

where $n = 0, 1, 2, \ldots$. Thereby he was able to explain in a brilliant manner the observed spectrum of the black-body radiation. Here, the constant $h =$

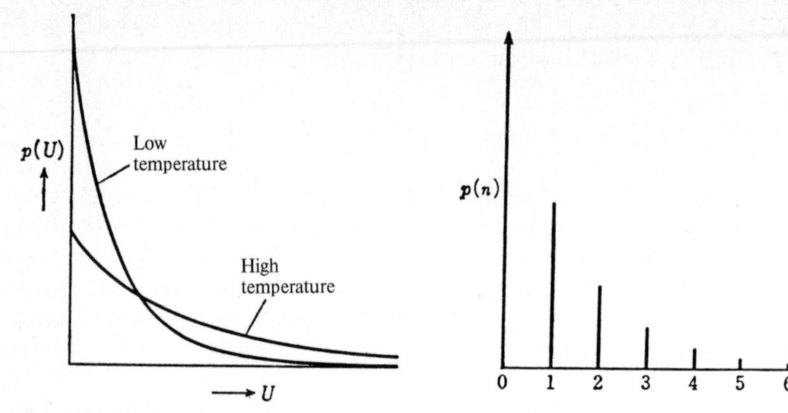

Fig. 4.2. Energy distributions at thermal equilibrium given by (4.7)

Fig. 4.3. Energy distribution at thermal equilibrium when $U = n\hbar\omega$

$2\pi\hbar = 6.626 \times 10^{-34}$ Js is now called the Planck constant and (4.8) is considered to represent the state in which there are n photons each having a unit energy of $\hbar\omega$. If the energy of electromagnetic waves can only have the discrete values given by (4.8), the energy distribution for a canonical distribution at thermal equilibrium will be as shown in Fig. 4.3. The average energy W_{th} in this case can be calculated to be

$$W_{\text{th}} = \frac{\sum_{n=0}^{\infty} n\hbar\omega \exp(-n\hbar\omega/k_{\text{B}}T)}{\sum_{n=0}^{\infty} \exp(-n\hbar\omega/k_{\text{B}}T)}, \tag{4.9}$$

which gives the thermal radiation of the mode with frequency ω. If we put

$$\exp(-\hbar\omega/k_{\text{B}}T) = r,$$

the series in the denominator of (4.9) becomes

$$\sum_{n=0}^{\infty} r^n = \frac{1}{1-r},$$

and the series in the numerator becomes

$$\sum_{n=0}^{\infty} nr^n = r\frac{\partial}{\partial r}\sum_{n=0}^{\infty} r^n = \frac{r}{(1-r)^2},$$

so that we obtain

$$W_{\text{th}} = \hbar\omega\frac{r}{1-r} = \frac{\hbar\omega}{\exp(\hbar\omega/k_{\text{B}}T) - 1}. \tag{4.10}$$

On looking at the above equation, we find that we have $W_{th} \cong k_B T$ when $\hbar\omega \ll k_B T$, which is the same as the case when U was assumed to take continuous values. In the high-frequency range, where $\hbar\omega \gg k_B T$, however, W_{th} approaches 0 as the frequency increases, being markedly different from the case when U takes continuous values.

Equation (4.10) is the average energy of each mode of electromagnetic waves in thermal equilibrium with a body at temperature T. Since the mode density of electromagnetic waves with frequencies between ω and $\omega + d\omega$ is given by (4.5), the energy per unit volume of thermal radiation for all of the modes, i.e., the energy density of thermal radiation with circular frequencies between ω and $\omega + d\omega$ is obtained from (4.5, 10)

$$W_{th}(\omega)\,d\omega = \frac{\omega^2}{\pi^2 c^3} \cdot \frac{\hbar\omega\,d\omega}{\exp(\hbar\omega/k_B T) - 1} \,. \tag{4.11}$$

In terms of linear frequency ν this is rewritten as

$$W_{th}(\nu)\,d\nu = \frac{8\pi\nu^2}{c^3} \cdot \frac{h\nu\,d\nu}{\exp(h\nu/k_B T) - 1} \,. \tag{4.12}$$

This is called Planck's law of black-body radiation or Planck's equation of thermal radiation. The spectral intensity or the spectral profile of black-body radiation is shown for several temperatures in Fig. 4.4.

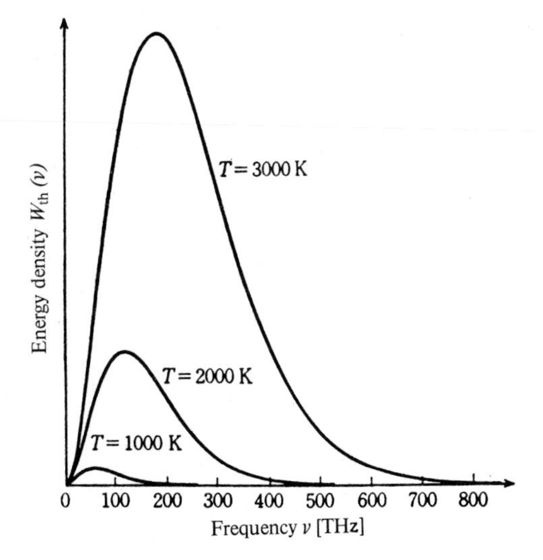

Fig. 4.4. Spectral distribution for black-body radiation (calculated)

4.3 Spontaneous Emission and Stimulated Emission

When a transition takes place in atoms between two energy levels, W_U and W_L, atoms in the upper level W_U emit light while atoms in the lower level W_L absorb light. The frequency of this light is given by the Bohr relation

$$\omega = \frac{W_U - W_L}{\hbar} . \tag{4.13}$$

This is true not only for atoms but also for molecules, ions, radicals, and atomic nuclei; however, in Chaps. 4 and 5 they will be treated in common under the general term as atoms.

Absorption of light by atoms in the lower level takes place in proportion to the intensity of the incident light. On the other hand, emission of light from atoms in the upper level takes place even in the absence of incident light; this is called spontaneous emission[1]. Let A be the probability of spontaneous emission from a single atom in unit time. Then the probability that an atom in the upper level emits light, when the energy density of the incident light is $W(\omega)$, is given by

$$p(\mathrm{U} \rightarrow \mathrm{L}) = A_{UL} + B_{UL} W(\omega) . \tag{4.14}$$

Here, the second term on the right-hand side represents the emission which is proportional to the intensity of the incident light and is called induced emission or stimulated emission. The probability that an atom in the lower level absorbs light is given by

$$p(\mathrm{L} \rightarrow \mathrm{U}) = B_{LU} W(\omega) . \tag{4.15}$$

If the upper and lower levels correspond, respectively, to single eigenstates[2], we have

$$B_{UL} = B_{LU} . \tag{4.16}$$

If the eigenstate of the upper level is g_U-fold degenerate and the lower level is g_L-fold degenerate, the total transition probability, taking into account the degeneracy, is given by

$$g_U B_{UL} = g_L B_{LU} ,$$

since the relation (4.16) holds for transitions between each pair of eigenstates. Unless stated otherwise, we shall consider transitions between nondegenerate levels hereafter. If either or both levels are degenerate, the corresponding results can be easily obtained by the superposition of transitions between all sublevels.

1 Sometimes called spontaneous radiation.
2 Quantum-mechanical proper states without degeneracy.

Since (4.14, 15) were derived by *A. Einstein* in 1916, coefficients A_{UL} and B_{UL} are called the Einstein A and B coefficients. When it is not necessary to indicate the upper and lower levels, they are simply written as A and B without any suffix.

Suppose that there are N_L atoms in the lower level. The light energy absorbed by these atoms in unit time, namely, the absorbed power, is given by

$$P_{abs} = \hbar\omega BW(\omega)\, N_L \,. \tag{4.17}$$

On the other hand, the light energy emitted in unit time from N_U atoms in the upper level, namely, the emitted power, is given by

$$P_{emi} = \hbar\omega\,[A + BW(\omega)]\, N_U \,. \tag{4.18}$$

When an atomic system is in thermal equilibrium at temperature T, the Boltzmann distribution or canonical distribution will hold, so that

$$N_U = N_L \exp\left(-\frac{\hbar\omega}{k_B T}\right) . \tag{4.19}$$

Now, such an atomic system is in equilibrium with the black-body radiation. Consequently, in a thermal radiation field with an energy density of $W_{th}(\omega)$, the emitted power and the absorbed power are equal. Thus we put $P_{abs} = P_{emi}$ when $W(\omega) = W_{th}(\omega)$, and we obtain from (4.17, 18),

$$W_{th}(\omega) = \frac{A}{B} \cdot \frac{N_U}{N_L - N_U} . \tag{4.20}$$

Substitution of (4.19) gives

$$W_{th}(\omega) = \frac{A}{B} \cdot \frac{1}{\exp(\hbar\omega/k_B T) - 1} . \tag{4.21}$$

According to quantum mechanics, when there are n photons in one mode, the probability for an atom in the upper level emitting a photon is written as

$$p(\mathrm{U} \rightarrow \mathrm{L}) = (n + 1)A \,,$$

while the probability for an atom in the lower level absorbing a photon is written as

$$p(\mathrm{L} \rightarrow \mathrm{U}) = nA \,.$$

Now, if the mode density is $m(\omega)$, there are $m(\omega)\,d\omega$ modes within the frequency range between ω and $\omega + d\omega$. Supposing that there are n photons on average in each mode[3], the energy density of radiation is given by

$$W(\omega) = m(\omega)\, n\hbar\omega \,.$$

---

3 As the electromagnetic field of each mode can be represented by a harmonic oscillator, its energy is $(n + 1/2)\,\hbar\omega$ where n is an integer. The term $\hbar\omega/2$ is the zero-point energy and gives the quantum fluctuation but it cannot be detected in any optical energy measurements.

Using this to rewrite the two equations above, we have

$$p(\mathrm{U} \to \mathrm{L}) = \frac{A}{m(\omega)\hbar\omega} W(\omega) + A ,$$

$$p(\mathrm{L} \to \mathrm{U}) = \frac{A}{m(\omega)\hbar\omega} W(\omega) .$$

By comparing these with (4.14, 15), it is immediately seen that

$$B = \frac{A}{m(\omega)\hbar\omega} . \tag{4.22}$$

Using the mode density (4.5), we find

$$\frac{A}{B} = m(\omega)\hbar\omega = \frac{\hbar\omega^3}{\pi^2 c^3} .$$

This is substituted into (4.21) to obtain

$$W_{\mathrm{th}}(\omega) = \frac{\hbar\omega^3}{\pi^2 c^3} \cdot \frac{1}{\exp(\hbar\omega/k_{\mathrm{B}}T) - 1} , \tag{4.23}$$

which is exactly the same as Planck's equation of thermal radiation (4.11) previously obtained.

Whereas absorption and induced emission of radiation take place only in the one mode of the incident light, spontaneous emission takes place in all of the $m(\omega)\,d\omega$ modes lying within the spectral width $d\omega$, so that we have $A = m(\omega)\, B\hbar\omega$. The additional factor $\hbar\omega$ here is due to the fact that, in (4.14, 15), the emission and absorption probabilities were expressed in terms of the energy density $W(\omega)$ of radiation rather than in terms of the number of photons. If, instead of $W(\omega)$, we use the number density of photons $\varrho(\omega)$ and express the emission probability as $A' + B'\varrho(\omega)$, we will obtain $A' = m(\omega) B'$. So far, we have only considered the ratio of A to B; the absolute values of A and B will be treated in the next chapter.

Now, if we neglect all but the two energy levels W_{U} and W_{L} of the atoms in thermal equilibrium, we will only find monochromatic light of the frequency given by the Bohr relation (4.13). At other frequencies there is no interaction between the radiation and this two-level atom so that $A = B = 0$, and hence the atomic system is perfectly transparent at these frequencies. A black body, on the other hand, absorbs and emits light of all frequencies, since its energy levels are distributed continuously. Even when the energy levels are distributed continuously, the number of atoms at thermal equilibrium in an upper level which is exactly $\hbar\omega$ above a lower level is equal to $\exp(-\hbar\omega/k_{\mathrm{B}}T)$ times the number of atoms in the lower level. Thus, for a black body, (4.23) holds at any arbitrary frequency, i.e., it expresses the black-body radiation as a continuous function of ω.

In Sect. 4.2 the equation of thermal radiation was derived on the assumption that the *photons* were distributed in a canonical distribution. In contrast, the equation of thermal radiation has been derived here on the premise that the photons were in thermal equilibrium with *atoms* with a Boltzmann distribution, and we arrived at exactly the same result. The fact that Planck's equation of thermal radiation agrees well with experimental results indicates not only that light energy is quantized but also that both spontaneous and induced emission take place in the emission of light from an excited atom. If A were zero, (4.20 or 21) would give $W_{\mathrm{th}} = 0$, and there would be no thermal radiation. Alternatively if we had $B_{\mathrm{UL}} = B_{\mathrm{LU}} = 0$, or $B_{\mathrm{UL}} = 0$ with $B_{\mathrm{LU}} \neq 0$, the spectrum of thermal radiation would be entirely different.

4.4 Dipole Radiation and Probability of Spontaneous Emission

According to the classical theory of electromagnetism, electromagnetic radiation is emitted when an electric charge is accelerated. Thus, an electric charge executing simple harmonic motion emits electromagnetic radiation at the frequency of oscillation. The magnitude of the moment of a dipole consisting of positive and negative charges $\pm e$ separated by a distance z is equal to ez, and its direction is from the charge $-e$ to the charge $+e$. Now, if the charges are in simple harmonic motion as given by

$$z = z_0 \mathrm{e}^{\mathrm{i}\omega t} + z_0^* \mathrm{e}^{-\mathrm{i}\omega t} , \tag{4.24}$$

the oscillating dipole moment $\mu(t)$ is

$$\mu(t) = p_0 \mathrm{e}^{\mathrm{i}\omega t} + p_0^* \mathrm{e}^{-\mathrm{i}\omega t} \tag{4.25}$$

where $p_0 = ez_0$.

The electromagnetic radiation emitted from this oscillating dipole is given by Maxwell's equations; the electric and magnetic fields at a distance r from the dipole are

$$\begin{aligned} E_\theta &= -\frac{p_0}{4\pi\varepsilon_0} \cdot \frac{k^2 \sin\theta}{r} \mathrm{e}^{\mathrm{i}(\omega t - kr)} + \text{c.c.} , \\ H_\varphi &= -\frac{\omega p_0}{4\pi} \cdot \frac{k \sin\theta}{r} \mathrm{e}^{\mathrm{i}(\omega t - kr)} + \text{c.c.} , \\ E_r &= E_\varphi = H_\theta = H_r = 0 , \end{aligned} \tag{4.26}$$

where $k = \omega/c$, φ is the azimuthal angle about the z axis, and c.c. stands for the complex conjugate of the preceding term. Here, only the fields at large distances are given, where $kr \gg 1$. The time average of the radial component of the Poynting vector $E_\theta H_\varphi$ becomes

$$P(\theta) = \frac{\omega^4|p_0|^2}{8\pi^2\varepsilon_0 c^3} \cdot \frac{\sin^2\theta}{r^2} . \tag{4.27}$$

Its integration over the entire solid angle gives the total power radiated from the oscillating dipole (4.25) as

$$P = \frac{\omega^4|p_0|^2}{3\pi\varepsilon_0 c^3} . \tag{4.28}$$

Now, the displacement z is quantum-mechanically represented by an operator. Consequently, the dipole moment $\mu = ez$ arising from the displacement of charges $\pm e$ is also represented by an operator. The dipole moment operator responsible for a transition between the upper and lower states is expressed by a matrix of two rows and two columns acting on the eigenfunctions of the upper (U) and lower (L) states. When the state is nondegenerate, the symmetry of the wave functions is fixed:

$$\psi_i(\boldsymbol{r}) = \psi_i(-\boldsymbol{r}) \text{ (even) }, \quad \text{or}$$
$$\psi_i(\boldsymbol{r}) = -\psi_i(-\boldsymbol{r}) \text{ (odd) },$$

where $i = \mathrm{U, L}$. Then the diagonal elements of the dipole moment matrix of a two-level atom are 0, $\mu_{\mathrm{UU}} = \mu_{\mathrm{LL}} = 0$. This is because ez is an odd function and $\psi_i^* ez\, \psi_i$ is an odd function whichever ψ_i is even or odd, so that

$$\mu_{ii} = \int \psi_i^*(\boldsymbol{r})\, ez\, \psi_i(\boldsymbol{r})\, d\boldsymbol{r} = 0 . \tag{4.29}$$

The off-diagonal elements μ_{UL} and μ_{LU} are not zero when the symmetries of the upper and lower states are different, and they represent the probability amplitude of the transition. Since all operators representing physical quantities are Hermitian operators, we have

$$\mu_{\mathrm{UL}} = \mu_{\mathrm{LU}}^* . \tag{4.30}$$

According to quantum mechanics the probability that an atom in the upper level with such a dipole moment emits a photon in unit time is given by

$$A = \frac{\omega^3}{3\pi\varepsilon_0\hbar c^3}|\mu_{\mathrm{UL}}|^2 . \tag{4.31}$$

Therefore, the radiation power of spontaneous emission from an atom in the upper level, $P_{\mathrm{emi}} = A\hbar\omega$, becomes

$$P_{\mathrm{emi}} = \frac{\omega^4}{3\pi\varepsilon_0 c^3}|\mu_{\mathrm{UL}}|^2 . \tag{4.32}$$

Comparing this with (4.28), which was obtained from the classical theory of electromagnetism, it becomes exactly the same if we assume $|p_0| = |\mu_{UL}|$. This shows that the process of quantum-mechanical spontaneous emission may be thought of in terms of an equivalent classical dipole radiation. Not only are the Maxwell equations relativistically correct, but they also describe correctly the behavior of a single photon based on quantum mechanics. However, they are not able to explain those phenomena which require second quantization, i.e., the quantum phenomena in which eigenstates for n photons are involved.

The wave function in a Schrödinger equation is generally considered to change with time. The Schrödinger representation of the wave function $\psi_n(\boldsymbol{r}, t)$ may be separated into a temporal function and a spatial function $\phi_n(\boldsymbol{r})$ in the form

$$\psi_n(\boldsymbol{r}, t) = \exp[-\mathrm{i}(W_n/\hbar)t]\,\phi_n(\boldsymbol{r}) \tag{4.33}$$

where W_n is the eigenenergy of state n. In contrast to the Schrödinger representation, the Heisenberg representation of the wave function does not change with time; it is instead the physical quantity which changes with time. Here, we shall use the Schrödinger representation.

The non-diagonal matrix elements of the z component of the transition dipole moment between the upper and the lower levels, U and L, can be expressed as

$$\begin{aligned} \int \psi_{\mathrm{U}}^{*}\, ez\, \psi_{\mathrm{L}}\, d\boldsymbol{r} &= \int \phi_{\mathrm{U}}^{*}\, ez\, \phi_{\mathrm{L}}\, d\boldsymbol{r}\, \mathrm{e}^{\mathrm{i}\omega_0 t} = \mu_{\mathrm{UL}}\, \mathrm{e}^{\mathrm{i}\omega_0 t}\,, \\ \int \psi_{\mathrm{L}}^{*}\, ez\, \psi_{\mathrm{U}}\, d\boldsymbol{r} &= \int \phi_{\mathrm{L}}^{*}\, ez\, \phi_{\mathrm{U}}\, d\boldsymbol{r}\, \mathrm{e}^{-\mathrm{i}\omega_0 t} = \mu_{\mathrm{LU}}\, \mathrm{e}^{-\mathrm{i}\omega_0 t}\,, \end{aligned} \tag{4.34}$$

where e is the electron charge. It is seen that, with $\omega_0 = (W_{\mathrm{U}} - W_{\mathrm{L}})/\hbar$, these equations correspond, respectively, to the first and second terms of the right-hand side of (4.25) for the classical dipole. Thus, by putting $p_0 = \mu_{\mathrm{UL}}$, we are able to obtain the probability of spontaneous emission from the calculation of classical dipole radiation.

Now that we have obtained the probability of spontaneous emission (4.31), we can obtain the Einstein B coefficient, which gives the probabilities of induced emission and absorption. By using (4.22), which was obtained from the calculation of the number of modes of electromagnetic waves, we obtain, for incident light of frequency ω,

$$B = \frac{\pi}{3\,\varepsilon_0 \hbar^2}|\mu_{\mathrm{UL}}|^2\,. \tag{4.35}$$

It must, however, be noted that this has been derived for the case where the incident light is unpolarized with respect to the direction of the atomic dipole moment. Equation (4.35) is also applicable when the atomic dipole moment is in a random direction with respect to the polarization of the incident light, so that the x, y, and z components μ_x, μ_y, and μ_z are related by

$$\langle|\mu_x|^2\rangle = \langle|\mu_y|^2\rangle = \langle|\mu_z|^2\rangle = \tfrac{1}{3}|\mu_{\rm UL}|^2 ,$$

where $\langle\ \ \rangle$ denotes the statistical average and the suffix UL has been omitted. On the other hand, if the atomic dipole is parallel to the polarization of the incident light, the coefficient of induced emission and absorption is given by

$$B = \frac{\pi}{\varepsilon_0\hbar^2}|\mu_{\rm UL}|^2 . \tag{4.36}$$

This B coefficient has been expressed in terms of circular frequency. In terms of linear frequency, on the other hand, a different result is obtained, as seen below.

The Einstein A coefficient representing spontaneous emission at frequency ν is obtained from (4.31) by substituting $2\pi\nu$ for ω as

$$A = \frac{16\pi^3\nu^3}{3\,\varepsilon_0 hc^3}|\mu_{\rm UL}|^2 . \tag{4.37}$$

However, the B coefficient is different from that obtained by merely putting $\omega = 2\pi\nu$ in (4.35 or 36). This is because the number of modes $m(\nu)\,d\nu$ in the frequency range between ν and $\nu + d\nu$ is different from the number of modes $m(\omega)\,d\omega$ in the circular frequency range between ω and $\omega + d\omega$, as explained in Sect. 4.1. If we express B in terms of ν instead of ω, (4.22) becomes

$$B_\nu = \frac{A}{m(\nu)\,h\nu}$$

and it can be rewritten by substituting (4.6 and 37):

$$B_\nu = \frac{2\pi^2}{3\,\varepsilon_0 h^2}|\mu_{\rm UL}|^2 . \tag{4.38}$$

The suffix ν indicates that the energy density of the incident light is expressed in terms of linear frequency instead of circular frequency.

Both (4.35 and 38) are averages for an atom which is directed at random with respect to the direction of polarization of the incident light. If the dipole moments of all the atoms are aligned in the same direction as the polarization of the incident light, the factor 3 in the denominator will be absent. In general, we should write $\langle|\mu_z|^2\rangle$ rather than $(1/3)\,|\mu_{\rm UL}|^2$, when the polarization of the incident light is in the z direction. If all the atoms are aligned in the z direction we have $\langle|\mu_z|^2\rangle = |\mu_{\rm UL}|^2$, while if they are in completely random directions we have a factor 1/3. Any intermediate case can be evaluated by using $\langle|\mu_z|^2\rangle$. Here, (4.38) represents the most typical case for randomly oriented dipoles.

4.5 Absorption of Light

If the intensity of light incident on a body in thermal equilibrium at a certain temperature T is higher than that of the black-body radiation at that temperature, the incident light is absorbed and attenuated. Since the absorber is not, in general, a black body, and the spectral distribution of the incident light is different from that of a black body, we shall treat each frequency component separately. If the energy density $W(\omega)$ of the incident light at frequency ω is greater than that of the thermal radiation $W_{\mathrm{th}}(\omega)$ of the absorber, the power absorbed by the absorber (4.17) is greater than the emission power (4.18), resulting in net absorption. Therefore, if a bright light source is placed on the far side of an object, an absorption spectrum of the object will be observed. On the other hand, if the far side of the object is dark (that is to say, the background is at a lower temperature than the object), an emission spectrum of the object will be observed. If the object is a perfect absorber with no transmission, only the thermal radiation of the object will be observed.

We shall now suppose that the thermal radiation of the object is negligible and investigate only the absorption of strong incident light. Unlike the thermal radiation, which has a continuous spectrum, the incident light is supposed to be almost perfectly monochromatic, as from a laser. In other words, no matter how narrow the width of the absorption line of the absorber may be, the spectral width of the light source is supposed to be much narrower. Denoting the frequency of the light source by ω, we shall express the frequency distribution of the coefficient of induced transition in the neighborhood of the central frequency ω_0 by

$$B(\omega) = Bg(\omega) . \tag{4.39}$$

Since the Einstein B coefficient is defined for the case where the spectral distribution of the light source is broader than that of the spectral line, it follows that

$$\int_0^\infty B(\omega)\, d\omega = B , \quad \int_0^\infty g(\omega)\, d\omega = 1 ,$$

where $g(\omega)$ represents the normalized spectral profile.

In terms of $B(\omega)$, the light power absorbed in a unit volume by an absorber with N_{U} and N_{L} atoms in the upper and lower levels, respectively, is given by

$$\Delta P = (N_{\mathrm{L}} - N_{\mathrm{U}})\hbar\omega B(\omega)\frac{P}{c} , \tag{4.40}$$

where P is the power of the incident light through a unit area. Thus, taking the z axis along the direction of propagation, we obtain

$$\frac{d}{dz}P(z) = -(N_{\mathrm{L}} - N_{\mathrm{U}})\frac{\hbar\omega}{c}B(\omega)P(z) .$$

The amplitude absorption constant $\alpha(\omega)$ is defined by

$$\frac{d}{dz}P(z) = -2\alpha(\omega)P(z) .$$

Comparing these two equations, we find the amplitude absorption constant to be

$$\alpha(\omega) = (N_L - N_U)\frac{\hbar\omega}{2c}B(\omega) . \tag{4.41}$$

Substitution of (4.39, 35) into (4.41) gives

$$\alpha(\omega) = (N_L - N_U)\frac{\pi\omega}{6\varepsilon_0\hbar c}|\mu_{UL}|^2 g(\omega) . \tag{4.42}$$

There are a variety of profiles for spectral lines, but the Lorentzian and the Gaussian profiles are the fundamental ones. The function $g_L(\omega)$ for the Lorentzian profile is defined by

$$g_L(\omega) = \frac{1}{\pi}\cdot\frac{\Delta\omega}{(\omega-\omega_0)^2 + (\Delta\omega)^2} . \tag{4.43}$$

When $\omega - \omega_0 = \pm\Delta\omega$, $g_L(\omega)$ becomes one half of the maximum value $g_L(\omega_0) = 1/\pi\Delta\omega$, and $\Delta\omega$ is called the halfwidth at half maximum (abbreviated to HWHM).

The Gaussian profile function having a HWHM of $\Delta\omega$ is given by

$$g_G(\omega) = \sqrt{\frac{\ln 2}{\pi}}\cdot\frac{1}{\Delta\omega}\exp\left[-\ln 2\cdot\left(\frac{\omega-\omega_0}{\Delta\omega}\right)^2\right] . \tag{4.44}$$

The maximum value at the center of the spectral line $\omega = \omega_0$ is

$$g_G(\omega_0) = \sqrt{\frac{\ln 2}{\pi}}\cdot\frac{1}{\Delta\omega} = \frac{0.470}{\Delta\omega} ,$$

which is $\sqrt{\pi\ln 2} = 1.476$ times $g_L(\omega_0)$ for the Lorentzian profile. Profiles of $g_L(\omega)$ and $g_G(\omega)$ with the same value of $\Delta\omega$ are compared in Fig. 4.5. As can be seen from the figure, the Lorentzian profile has a longer tail than the Gaussian profile, and therefore has a smaller maximum value as a result of normalization.

In expressing the absorption or induced emission of light by atoms, we suppose that each atom has a cross-section σ, and that the incident light falling within this area will be absorbed or produce induced emission; σ for the atom in the lower level is called the absorption cross-section and for the atom in the upper level is called the induced emission cross-section. Since the induced emission cross-section is equal to the absorption cross-section, as can be clearly seen from (4.16), they will both be denoted by $\sigma(\omega)$ for mono-

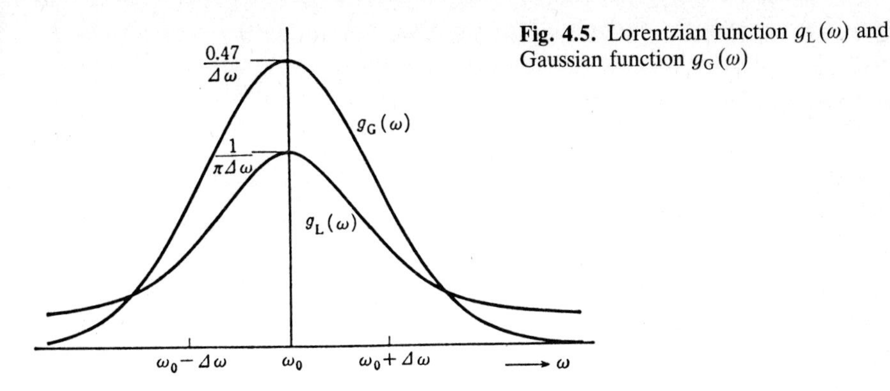

Fig. 4.5. Lorentzian function $g_L(\omega)$ and Gaussian function $g_G(\omega)$

chromatic light of frequency ω. Thus, the power absorption constant of a medium, in which there are N_U and N_L atoms per unit volume in the upper and lower levels, respectively, is given by

$$2\alpha(\omega) = (N_L - N_U)\,\sigma(\omega)\ . \tag{4.45}$$

Therefore, by comparing (4.45) with (4.42), we can express the absorption cross-section for the atom in this transition as

$$\sigma(\omega) = \frac{\pi\omega}{3\,\varepsilon_0\hbar c}\,|\mu_{UL}|^2 g(\omega)\ , \quad \text{or}$$

$$\sigma(\omega) = \frac{\hbar\omega}{c}\,B(\omega)\ . \tag{4.46}$$

Additionally, when the medium is in thermal equilibrium at temperature T, the relation (4.19) holds between N_U and N_L, so that we have

$$N_L - N_U = (1 - e^{-\hbar\omega/k_BT})\,N_L\ .$$

When $\hbar\omega \ll k_BT$ for low frequencies or high temperatures, we have approximately

$$N_L - N_U \simeq \frac{\hbar\omega}{k_BT}\,N_L\ ,$$

whereas for high frequencies (light) or low temperatures, when $\hbar\omega \gg k_BT$, we may approximate the relation as

$$N_L - N_U \simeq N_L\ .$$

4.6 Complex Susceptibility and Refractive Index

Absorption and emission of light are inevitably accompanied by dispersion. Thus these processes must be treated by considering the variation of the phase as well as the magnitude of the light amplitude and the energy. In Sect. 4.4 we stated that the transition between two energy levels of an atom could be treated by considering a classical oscillating dipole; similarly, we shall derive here the optical properties of a medium consisting of numerous atoms which are considered as classical oscillators.

The differential equation of the amplitude x of an oscillator with proper frequency ω_0 and damping constant γ, driven by an external force F, is given by

$$\frac{d^2x}{dt^2} + 2\gamma\frac{dx}{dt} + \omega_0^2 x = \frac{F}{m} . \tag{4.47}$$

Here, m is the mass of the oscillator. By denoting its charge by $-e$, the above equation, for the case when the driving electric field is the incident light $E(\omega)\exp(\mathrm{i}\omega t)$ becomes

$$\frac{d^2x}{dt^2} + 2\gamma\frac{dx}{dt} + \omega_0^2 x = -\frac{e}{m}E(\omega)\,\mathrm{e}^{\mathrm{i}\omega t} .$$

The steady-state solution can be written in the form

$$x = x(\omega)\,\mathrm{e}^{\mathrm{i}\omega t}$$

so that, by substituting this into the equation above, it can be easily found that

$$x(\omega) = \frac{e}{m}\cdot\frac{E(\omega)}{\omega^2 - 2\,\mathrm{i}\gamma\omega - \omega_0^2} , \tag{4.48}$$

which can be approximated to

$$x(\omega) = \frac{e}{2m\omega_0}\cdot\frac{E(\omega)}{\omega - \omega_0 - \mathrm{i}\gamma} , \tag{4.49}$$

when the damping is so small that $\gamma \ll \omega$.

When there are N_U and N_L atoms per unit volume in the medium in the upper and lower levels, respectively, the complex polarization of the medium, expressed in the form $P(\omega)\exp(\mathrm{i}\omega t)$, is given by

$$P(\omega) = -ex(\omega)(N_\mathrm{L} - N_\mathrm{U}) . \tag{4.50}$$

Since the complex susceptibility $\chi(\omega) = \chi'(\omega) - \mathrm{i}\chi''(\omega)$ is defined by

$$P(\omega) = \varepsilon_0\chi(\omega)E(\omega) ,$$

we obtain from (4.49, 50)

$$\chi(\omega) = -\frac{(N_\mathrm{L} - N_\mathrm{U})\,e^2}{2\,\varepsilon_0 m\omega_0}\cdot\frac{1}{\omega - \omega_0 - \mathrm{i}\gamma} . \tag{4.51}$$

Separating this into its real part $\chi'(\omega)$ and imaginary part $-\chi''(\omega)$, we have

$$\chi'(\omega) = -\frac{(N_L - N_U)e^2}{2\varepsilon_0 m\omega_0} \cdot \frac{\omega - \omega_0}{(\omega - \omega_0)^2 + \gamma^2}, \tag{4.52}$$

$$\chi''(\omega) = \frac{(N_L - N_U)e^2}{2\varepsilon_0 m\omega_0} \cdot \frac{\gamma}{(\omega - \omega_0)^2 + \gamma^2}. \tag{4.53}$$

Plotting these against ω, we obtain Fig. 4.6.

In general, when the complex susceptibility $\chi(\omega)$ is given, the complex dielectric constant becomes

$$\varepsilon(\omega) = \varepsilon_0[1 + \chi(\omega)].$$

Assuming the magnetic permeability of the medium to be $\mu = \mu_0$, the complex refractive index η is written as

$$\eta = \eta' - \mathrm{i}\kappa = \sqrt{\frac{\varepsilon(\omega)}{\varepsilon_0}} = \sqrt{1 + \chi(\omega)}, \tag{4.54}$$

where η' is the real part, and κ is the imaginary part, called the extinction coefficient. It is proportional to the absorption constant as shown in the following. The amplitude of a plane wave propagating in the z direction is expressed as $\exp(\mathrm{i}\omega t - \mathrm{i}kz)$. In a medium in which the refractive index is expressed by the complex quantity $\eta' - \mathrm{i}\kappa$, we have

$$k = (\eta' - \mathrm{i}\kappa)\frac{\omega}{c}. \tag{4.55}$$

Thus the amplitude

$$\exp(\mathrm{i}\omega t - \mathrm{i}kz) = \exp\left(-\frac{\kappa\omega}{c}z\right) \cdot \exp\left(\mathrm{i}\omega t - \mathrm{i}\eta'\frac{\omega}{c}z\right)$$

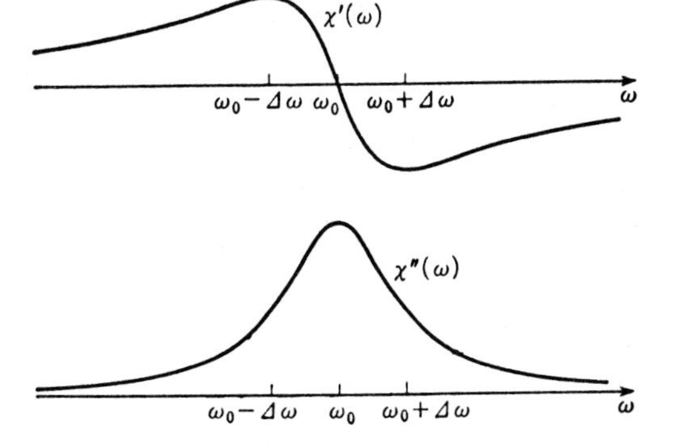

Fig. 4.6. Real part $\chi'(\omega)$ and imaginary part $-\chi''(\omega)$ of the complex susceptibility $\chi(\omega)$

indicates a decrease in the form $\exp(-\alpha z)$, so that the absorption constant is

$$\alpha = \frac{\omega}{c}\kappa . \tag{4.56}$$

Since κ is approximately equal to $\chi''/2$ when $|\chi(\omega)| \ll 1$ from (4.54), substitution of (4.53) into (4.56) gives

$$\alpha \simeq \frac{(N_L - N_U)\,e^2}{4\,\varepsilon_0 mc} \cdot \frac{\gamma}{(\omega - \omega_0)^2 + \gamma^2} , \tag{4.57}$$

showing that the absorption line shape is Lorentzian. If we replace the term e^2/mc in this equation by the quantum-mechanical term $2\omega\,|\mu_{UL}|^2/3\hbar$, we obtain the same result as the quantum-mechanical calculation of (4.42), assuming the Lorentzian profile (4.43). Thus, the real and imaginary parts of the complex susceptibility can be expressed as

$$\chi'(\omega) = -(N_L - N_U)\,\frac{|\mu_{UL}|^2}{3\,\varepsilon_0\hbar} \cdot \frac{\omega - \omega_0}{(\omega - \omega_0)^2 + \gamma^2} , \tag{4.58}$$

$$\chi''(\omega) = (N_L - N_U)\,\frac{|\mu_{UL}|^2}{3\,\varepsilon_0\hbar} \cdot \frac{\gamma}{(\omega - \omega_0)^2 + \gamma^2} . \tag{4.59}$$

Atoms in a normal medium are in thermal equilibrium or very nearly so; therefore, it follows that $N_L > N_U$. In this case $\chi'' > 0$ from (4.59), that is to say, the absorption constant α is positive and the medium absorbs light. If, however, we can make $N_L < N_U$, we have $\chi'' < 0$ and $\alpha < 0$, which means that the absorption is negative and that light can be amplified. Since a population distribution such that $N_L < N_U$ is contrary to that of thermal equilibrium, it is called population inversion in the terminology of laser physics. It will be discussed in greater detail in the next chapter.

In this chapter the complex susceptibility was derived and the absorption coefficient was obtained on the assumption that the medium was equivalent to an ensemble of classical oscillators, where the magnitude of the classical dipole corresponding to the atomic dipole was determined from the dipole radiation. A quantum-mechanical treatment of the complex susceptibility including the saturation effect will be deferred to Chap. 8.

Problems

4.1 Find the mode density of electromagnetic waves in a two-dimensional space between two plane-parallel perfect conductors. The separation between the two conductors may be assumed to be less than one half-wavelength.

Answer: $m(\omega)\,d\omega = \dfrac{\omega}{\pi c^2}\,d\omega$.

4.2 Rewrite Planck's equation for the energy density of thermal radiation as a function of the wavelength of the radiation.

Answer: $$W(\lambda)\,d\lambda = \frac{8\pi hc}{\lambda^5}\,\frac{d\lambda}{\exp(hc/\lambda k_B T) - 1}\,.$$

4.3 Find the pressure exerted by thermal radiation on the surface of a perfect reflector and a black body at zero temperature.

Answer: Using

$$\int_0^\infty \frac{x^3 dx}{e^x - 1} = \frac{\pi^4}{15}\,,$$

we obtain $(8\pi^5 k^4/45\,c^3 h^3)\,T^4$ and $(4\pi^5 k^4/45\,c^3 h^3)\,T^4$, respectively. It may be noted that the recoil due to emission from the black body in equilibrium at temperature T is just equal to the radiation pressure.

4.4 The Einstein B coefficient has been defined by (4.14) in this text. Different definitions may often be found in other literature. Find the relations corresponding to (4.22) and (4.35) when B is defined by using the light intensity instead of the energy density. Note that the light intensity $Id\Omega$ in a solid angle $d\Omega$ is defined by the magnitude of the Poynting vector of the radiation within that solid angle.

Answer: Using $I = cW/4\pi$, we obtain

$$B = \frac{4\pi A}{cm(\omega)\,\hbar\omega}\,, \quad \text{and} \quad B = \frac{4\pi^2}{3\,\varepsilon_0\hbar^2 c}\,|\mu_{\mathrm{UL}}|^2$$

respectively. It may be remarked that the latter becomes

$$B = \frac{16\pi^3}{3\hbar^2 c}\,|\mu_{\mathrm{UL}}|^2$$

in the cgs system of units.

4.5 Verify that (4.26) is a solution of Maxwell's equations.

Hint: Rewrite Maxwell's equations in spherical coordinates.

4.6 Derive (4.28) from (4.27).

4.7 Show that the total energy emitted by a classical dipole as given by (4.25) will be $\hbar\omega$ if $|p_0|^2$ decays from its initial value $|\mu_{\mathrm{UL}}|^2$ at a rate equal to the Einstein A coefficient.

4.8 Show that the power spectrum of a damped oscillation is given by a Lorentzian profile function.

Hint: Use (2.17, 19).

4.9 Calculate the absorption cross-section of an atomic transition in the ideal case when it is broadened only by its lifetime such that $\Delta\omega = A$.

Answer: $\sigma = \lambda^2/4\pi$.

4.10 Draw the locus representing the complex susceptibility (4.51) on a complex plane, as ω varies from 0 to infinity.

Answer: A circle passing through the origin, centered at $-\mathrm{i}(N_\mathrm{L} - N_\mathrm{U})\, e^2/4\,\varepsilon_0 m\omega_0\gamma$.

4.11 Find the optical depth of resonance radiation at a wavelength of 1 μm in a gas of atoms having a transition dipole moment of 1 debye and a Lorentzian linewidth of HWHM = 1 GHz, at a pressure of 1 Pa and temperature of 300 K. Note that the light intensity attenuates to 1/e of its initial value in passing the optical depth and that one debye is equal to 10^{-18} cgs-esu or 3.335×10^{-30} Coulomb-meter.

Answer: 1.04 mm.

5. Principle of the Laser

Unlike ordinary light sources such as lamps, electric bulbs, or discharge tubes, the laser is an oscillator similar to a radio transmitter. In this chapter an elementary theory of laser action is presented with the help of circuit theory and rate equations. The semiclassical theory and the quantum-mechanical theory of the laser are deferred to Chap. 9.

The basic structure of a laser consists of an amplifying medium with inverted population (to be explained below) between two mirrors M_1 and M_2, as shown in Fig. 5.1. The two mirrors constitute a Fabry-Perot resonator which confines light at a resonant frequency between the mirrors. Besides plane mirrors, concave mirrors, diffraction gratings, or Bragg reflectors are also used, while occasionally an optical resonator with more than two mirrors is used. As explained in Chap. 1, there are many kinds of amplifying media which can be used in the laser to play the essential role of amplifying light through population inversion and stimulated emission. There are, however, lasers such as the Raman laser which do not depend on population inversion.

5.1 Population Inversion

As explained in Sect. 4.3, stimulated emission and absorption occur simultaneously as long as the atoms are distributed in upper and lower levels. The probability of an induced transition from the upper level to the lower level is the same as that from the lower level to the upper level. In the normal state of the medium there are more atoms in the lower level than in the upper level, resulting in net absorption of light. If the medium is excited by an appropriate method so that the number of atoms in the upper level N_U is

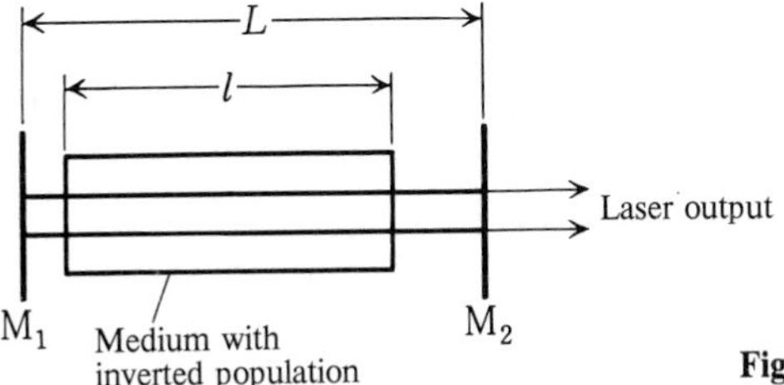

Fig. 5.1. Laser with a Fabry-Perot resonator

greater than the number in the lower level N_L, the light incident on the medium will be amplified by stimulated emission. This is laser amplification.

The process of making $N_U > N_L$, as opposed to $N_L > N_U$ in thermal equilibrium, is population inversion, while the atomic population with $N_U > N_L$ is the inverted population. It is customary, however, that both population inversion and inverted population refer to the inverted population *difference*. Now, if we formally apply (4.19) to the case of inverted population $N_U > N_L$, we obtain $\hbar\omega/k_B T < 0$, since $N_U/N_L = \exp(-\hbar\omega/k_B T) > 1$. As $\hbar\omega/k_B$ is a positive quantity, the inverted population corresponds to a negative temperature $T < 0$. Therefore, we may say that the state of inverted population has a negative temperature. Needless to say, this is not a temperature in the thermodynamical sense; it is nothing more than a parameter expressing the state of inverted population using (4.19).

In order to invert populations of atomic levels, the atoms must be excited by depositing energy in the medium using such method as to decrease the number N_L of atoms in the lower level and to increase the number N_U of atoms in the upper level. This process is called pumping, since the atoms are redistributed as if pumped from the lower level to the upper level. The commonly used methods of pumping are optical pumping, where the atoms are excited by illumination of light, excitation by electric discharge in the case of gases, and injection of carriers by a forward current through a *p-n* junction in the case of semiconductors. Besides these methods, there are others such as excitation by irradiation with electron beams or other radiations, excitation by chemical reaction or by shock waves, and so forth.

Formerly, before the advent of the maser, many believed that any process of population inversion was impossible. *V. A. Fabricant* is known to be the first person to start active research, in about 1940, to realize population inversion, although his efforts turned out to be unsuccessful [5.1]. After World War II, microwave spectroscopy was developed, and, in 1954, *C. H. Townes* et al. succeeded in realizing population inversion with a molecular beam of ammonia to make a maser of 1.25 cm wavelength [5.2]. As the ammonia molecules were distributed among energy levels in thermal equilibrium, the molecules in the upper level were collected and those in the lower level were eliminated by the action of an inhomogeneous electric field, so that population inversion could be achieved. However, such a method as this, where population inversion is established by decreasing the number of atoms in the lower level, cannot be applied to optical transitions.

The reason for this is as follows. We find from (4.19) that we have $N_U \approx N_L$ in the microwaves case, since $h\nu \ll k_B T$ at the microwave frequency ν. On the other hand, the population of the upper level N_U in the optical case is very small, since $h\nu \gg k_B T$ at the optical frequency ν. Therefore, in order to obtain stimulated emission in the optical region it is not sufficient merely to eliminate atoms in the lower level, but it is also necessary to increase the number of atoms in the upper level with a process of pumping. When two-level atoms are excited by irradiation or by electron collisions, the number of atoms in the upper level will increase, but at the same time the probability of de-excitation that brings these excited atoms back to the lower level will increase with the incident light or electrons. Consequently, no matter how strongly the atoms may be excited, population inversion cannot be obtained. Therefore, three or four levels of an atom must be used in order to obtain population inversion for laser action. It is not always necessary that the energy levels should be discrete and sharp, and band levels may be used in some cases. Thus the dye laser and the semiconductor laser can be considered basically as four-level lasers, to be described below.

5.2 Population Inversion in a Three-Level Laser

Besides the ruby laser mentioned in Sect. 1.2, there are many three-level lasers such as the optically pumped gas lasers. Let the energies and populations of the relevant three levels of the laser atom be denoted respectively by W_1, W_2, W_3, and N_1, N_2, N_3. If $W_3 > W_2 > W_1$, as shown in Fig. 5.2, we have $N_1 > N_2 > N_3$ in the three-level system in thermal equilibrium. Here the lowest state 1 is not necessarily the ground state of the atom. Atoms in level 1 will be excited to level 3 by collisions with photons, electrons, or excited atoms of appropriate energy. Without going into the process of excitation, we shall denote by Γ the probability of exciting the atom from level 1 to level 3 by any such method of pumping.

When the pumping is removed, the excited atoms will, in general, gradually return to the state of thermal equilibrium. This is known as relaxation. If we consider the atoms individually, the relaxation process takes place at the same time as other atoms are being excited. Besides the radiative process, where the excited atom makes a transition to the lower state by emitting a photon, there are non-radiative processes such as the collision of molecules in gases or the atom-lattice interaction in solids, where the excited atom makes a transition to the lower state by releasing its energy in the form of molecular kinetic energy or vibrational energy of the lattice. Since relaxation is the result of such statistical processes, we define the relaxation rate or the relaxation constant as a statistical average of the relaxation probabilities of

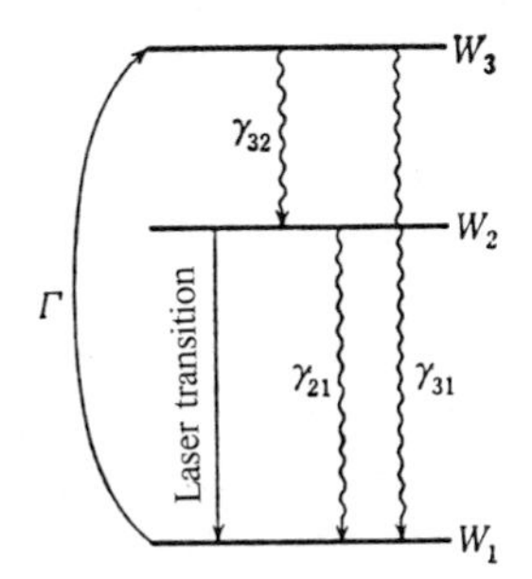

Fig. 5.2. Energy-level diagram of a three-level laser

the excited atoms in unit time. The reciprocal of the relaxation rate is the average lifetime of the excited atoms and, since it is equal to the decay time of fluorescence from the excited atoms, it is sometimes called the fluorescence lifetime.

Now, the probability[1] γ_{LU} of an atom being thermally excited from the lower state W_L to the upper state W_U is related to the probability γ_{UL} of the reverse process from W_U to W_L by thermal relaxation. This relation, in thermal equilibrium, is written as

$$N_U \gamma_{UL} = N_L \gamma_{LU}, \quad N_U = N_L \exp\left(-\frac{W_U - W_L}{k_B T}\right),$$

where T is the temperature of the medium. Then we find

$$\frac{\gamma_{LU}}{\gamma_{UL}} = \exp\left(-\frac{W_U - W_L}{k_B T}\right). \tag{5.1}$$

This last relation holds generally, even if N_U and N_L do not represent populations in thermal equilibrium.

If these probabilities are constant under the conditions considered, the rate equations expressing the rate of change of the number of atoms in each level of the three-level system under pumping are written

$$\frac{dN_1}{dt} = -(\Gamma + \gamma_{12} + \gamma_{13}) N_1 + \gamma_{21} N_2 + \gamma_{31} N_3 , \tag{5.2}$$

$$\frac{dN_2}{dt} = \gamma_{12} N_1 - (\gamma_{21} + \gamma_{23}) N_2 + \gamma_{32} N_3 , \tag{5.3}$$

$$\frac{dN_3}{dt} = (\Gamma + \gamma_{13}) N_1 + \gamma_{23} N_2 - (\gamma_{31} + \gamma_{32}) N_3 . \tag{5.4}$$

Here, $N_1 + N_2 + N_3 = \text{const} = N$, where N is the total number of atoms in the three-level system.

In the steady state, the distribution of the number of atoms under constant pumping can be obtained by putting the left-hand sides of (5.2–4) equal to zero. Although the solutions can be easily calculated, the expressions are lengthy. Therefore, we shall assume that the separations between the levels are sufficiently greater than the thermal energy $k_B T$, so that $\gamma_{12} \ll \gamma_{21}$, $\gamma_{13} \ll \gamma_{31}$, and $\gamma_{23} \ll \gamma_{32}$; thereby, we can neglect γ_{12}, γ_{13}, and γ_{23} in (5.2–4). Then, we obtain the steady-state solution:

$$N_1 = \frac{\gamma_{21}(\gamma_{31} + \gamma_{32})}{\gamma_{21}(\gamma_{31} + \gamma_{32}) + (\gamma_{21} + \gamma_{32})\Gamma} N , \tag{5.5}$$

$$N_2 = \frac{\gamma_{32}\Gamma}{\gamma_{21}(\gamma_{31} + \gamma_{32}) + (\gamma_{21} + \gamma_{32})\Gamma} N . \tag{5.6}$$

---

1 Short for probability rate per unit time.

If the excitation is so strong that

$$\Gamma > \gamma_{21}\left(1 + \frac{\gamma_{31}}{\gamma_{32}}\right), \tag{5.7}$$

we have $N_2 > N_1$, i.e., population inversion. In order to obtain population inversion with moderate pumping, γ_{21} should be small and γ_{32} should be large compared to γ_{31}, as can be seen from the condition above. In other words, it is desirable that the relaxation from the upper laser level to the lower laser level should be slow, while the relaxation from the uppermost level 3, to which the atom was initially excited, to the upper laser level 2 should be fast.

The population inversion as defined by $\Delta N = N_2 - N_1$ is calculated from (5.5, 6) as a function of the excitation intensity Γ to be

$$\Delta N = \frac{\gamma_{32}\Gamma - \gamma_{21}(\gamma_{31} + \gamma_{32})}{\gamma_{21}(\gamma_{31} + \gamma_{32}) + (\gamma_{21} + \gamma_{32})\Gamma} N \, . \tag{5.8}$$

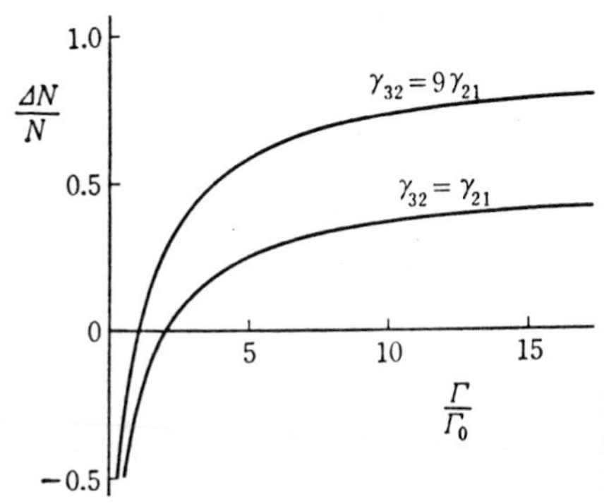

Fig. 5.3. Population inversion as a function of excitation intensity for a three-level system. $[\Gamma_0 = \gamma_{21}(\gamma_{31} + \gamma_{32})/(\gamma_{21} + \gamma_{32})]$

This is graphically shown in Fig. 5.3. The population inversion when the excitation is extremely strong is given by

$$\lim_{\Gamma \to \infty} \Delta N = \frac{\gamma_{32} N}{\gamma_{21} + \gamma_{32}} = \frac{N}{1 + \gamma_{21}/\gamma_{32}} \, . \tag{5.9}$$

Thus, we see from this also that the smaller γ_{21} and the larger γ_{32}, the greater the population inversion, resulting in a correspondingly stronger laser action. The threshold condition for laser oscillation and the output characteristics of the laser will be treated in Sect. 5.5 and in subsequent sections.

5.3 Population Inversion in a Four-Level Laser

Since the lower level of the laser transition is the lowest level in a three-level laser, the majority of the atoms ($N_1 \approx N$) are in this level at thermal equilibrium. Thus, in order to invert the population, the number of atoms in the lowest level 1 must be reduced to less than half by intense pumping. This requirement is much alleviated in a four-level system.

Let us consider an atom which has four energy levels as shown in Fig. 5.4; we want to invert the population between levels 2 and 1. If the lower laser level 1 lies at an energy higher than $k_B T$ above the ground level 0, the number of thermally excited atoms in the lower laser level 1 is so small that the population can be easily inverted by pumping a relatively small number of atoms into the upper level 2. The conditions for population inversion in this case are discussed below.

Although the separations between levels 1, 2, and 3 are assumed to be much greater than $k_B T$, as in the case of the three-level laser, we do not neglect the number of thermally excited atoms $\gamma_{01} N_0$ from the most populated ground level 0 to the nearest level 1. The rate equations for atomic populations in the four levels, then, become

$$\begin{aligned} \frac{dN_1}{dt} &= \gamma_{01} N_0 - \gamma_{10} N_1 + \gamma_{21} N_2 + \gamma_{31} N_3 \,, \\ \frac{dN_2}{dt} &= -\gamma_2 N_2 + \gamma_{32} N_3 \,, \\ \frac{dN_3}{dt} &= \Gamma N_0 - \gamma_3 N_3 \,, \\ -\frac{dN_0}{dt} &= \frac{dN_1}{dt} + \frac{dN_2}{dt} + \frac{dN_3}{dt} \,. \end{aligned} \tag{5.10}$$

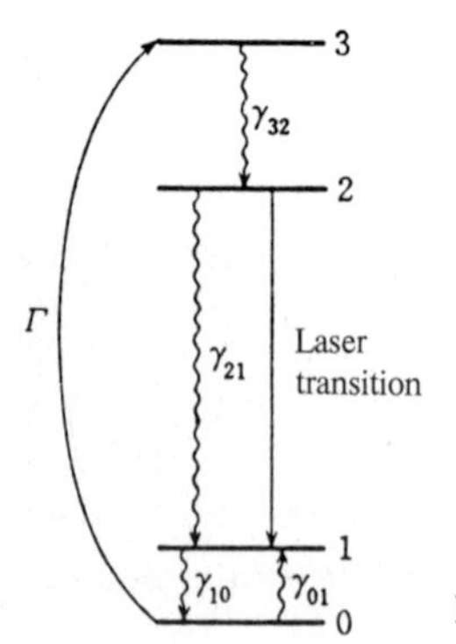

Fig. 5.4. Energy-level diagram of a four-level laser

The meanings of the coefficients in the above equations are the same as in the three-level system, except that we have put $\gamma_2 = \gamma_{20} + \gamma_{21}$ and $\gamma_3 = \gamma_{30} + \gamma_{31} + \gamma_{32}$.

The steady-state solution is obtained as before:

$$N_1 = \left(\frac{\gamma_{01}}{\gamma_{10}} + \frac{\gamma_{21}\gamma_{32} + \gamma_2\gamma_{31}}{\gamma_{10}\gamma_2\gamma_3}\Gamma\right) N_0 \,, \tag{5.11}$$

$$N_2 = \frac{\gamma_{32}\Gamma}{\gamma_2\gamma_3} N_0 \,, \tag{5.12}$$

$$N_3 = \frac{\Gamma}{\gamma_3} N_0 \,. \tag{5.13}$$

Since $N_0 + N_1 + N_2 + N_3 = N$, N_0 is given by

$$N_0 = \frac{\gamma_{10}\gamma_2\gamma_3 N}{\gamma_{01}\gamma_2\gamma_3 + \gamma_{32}(\gamma_{21} + \gamma_{10})\Gamma + \gamma_2(\gamma_{31} + \gamma_{10})\Gamma} \,. \tag{5.14}$$

From (5.11, 12), the condition for population inversion can be expressed as

$$\Gamma > \frac{\gamma_{01}\gamma_2\gamma_3}{\gamma_{32}\gamma_{10} - \gamma_{21}\gamma_{32} - \gamma_2\gamma_{31}} \,. \tag{5.15}$$

Now, γ_{01} in the numerator of this equation is the probability of thermal excitation from level 0 to level 1, and is a small quantity, as shown by the relation $\gamma_{01} = \gamma_{10} \exp(-W_1/k_B T)$; therefore, the excitation intensity Γ necessary for population inversion is lowered. Since $\gamma_{31} < \gamma_3 = \gamma_{31} + \gamma_{30} + \gamma_{32}$ and $\gamma_{21} < \gamma_2 = \gamma_{21} + \gamma_{20}$, (5.15) can be approximated by

$$\Gamma > \frac{\gamma_{01}\gamma_2\gamma_3}{\gamma_{10}\gamma_{32}} = \exp(-W_1/k_B T)\,\gamma_2\left(1 + \frac{\gamma_{31} + \gamma_{30}}{\gamma_{32}}\right) \tag{5.16}$$

when $\gamma_{10} \gg \gamma_2$. Comparing this with the condition (5.7) for population inversion in a three-level system, we see that they are very similar to one another, except for the factor $\exp(-W_1/k_B T)$. Since the four-level system has an extra level 0, it is quite obvious that we have $\gamma_{21} + \gamma_{20}$ instead of γ_{21}, and $\gamma_{31} + \gamma_{30}$ instead of γ_{31}. Here, it is the factor $\exp(-W_1/k_B T)$ that is important, because population inversion can be obtained even with very weak pumping if the lower laser level 1 is above the ground level 0 by at least a few times $k_B T$ in energy.

5.4 Laser Amplification

We shall consider here the reason why laser amplification and laser oscillation occur when the population is inverted in the medium. Since $N_U > N_L$ in the inverted population, χ'' is negative because the first factor $(N_L - N_U)$ on the right-hand side of (4.59) becomes negative. Therefore, the absorption constant α is negative so that $\exp(-\alpha z)$ increases with z instead of decreasing; in other words, $\alpha < 0$ for an inverted population means that amplification rather than absorption occurs. Since the power is proportional to the square of the amplitude, the power amplification or the gain in the medium of length z is $\exp(-2\alpha z) = \exp(Gz)$, where $G = -2\alpha$ is called the gain constant and $G/2$ the amplification constant[2].

The gain constant of the laser medium, whose population inversion is $\Delta N = N_2 - N_1$ between the upper level 2 and the lower level 1, is readily obtained from (4.41) as

$$G = \Delta N \frac{\hbar\omega}{c} B(\omega) , \tag{5.17}$$

which may be rewritten as

$$G = \Delta N \sigma(\omega) .$$

The upper and lower levels of the transition used in most lasers and masers are often degenerate. If an electric or magnetic field is applied to the atom, the degeneracy is partly or wholly removed to disclose a number of separate sublevels. Then we must consider each of the sublevels when calculating the thermodynamic equilibrium and the probability of induced transitions due to incident light. Thus, denoting by N_2 the *total* number of atoms in the upper level of degeneracy g_2, and by N_1 the *total* number of atoms in the lower level of degeneracy g_1, we have

$$\frac{N_2}{g_2} = \frac{N_1}{g_1} \exp\left(-\frac{\hbar\omega}{k_B T}\right) , \tag{5.18}$$

since the condition of thermal equilibrium should be applied to each of the sublevels.

If $g_2 > g_1$, it is possible for N_2 to be greater than N_1 in thermal equilibrium. Therefore, when there is any degeneracy, it is not appropriate to consider $N_2 - N_1$ as the population inversion. Nevertheless, if the number of atoms in each sublevel of the degenerate upper and lower levels were

2 To avoid confusion $G/2$ is sometimes called the amplitude amplification constant. It should be noted, however, that "ampli-" of amplification stands for amplitude in the first place.

denoted by N_2 and N_1 respectively, we might give the population inversion as $\Delta N = N_2 - N_1$. However, since N_1 and N_2 are customarily the total numbers of atoms when the levels are degenerate, the population inversion must be represented by

$$\Delta N = \frac{N_2}{g_2} - \frac{N_1}{g_1} \,. \tag{5.19}$$

Again, when considering induced emission and absorption, the transitions between the upper and lower sublevels are observed in multiples of g_2 and g_1, respectively, so that the Einstein B coefficient becomes

$$B = g_2 B_{21} = g_1 B_{12} \,, \tag{5.20}$$

as mentioned in Sect. 4.3.

Here, we have obtained the thermal equilibrium distribution (5.18) and the B coefficient (5.20) for the degenerate upper and lower levels of the laser transition, but these equations hold generally for any pair of levels.

The gain constant of the laser medium, whose population inversion is given by (5.19) and B coefficient by (5.20), can now be written as

$$G = \left(N_2 - \frac{g_2}{g_1} N_1\right) \frac{\pi\omega}{3\,\varepsilon_0 c\hbar} |\mu_{21}|^2 g(\omega) \,, \quad \text{or} \tag{5.21}$$

$$G = \left(\frac{g_1}{g_2} N_2 - N_1\right) \frac{\pi\omega}{3\,\varepsilon_0 c\hbar} |\mu_{12}|^2 g(\omega) \,,$$

by substituting (5.19, 4.39, 35) into (5.17).

Laser amplification in a medium with population inversion is the reverse process of absorption in an ordinary medium at thermal equilibrium. The incident light is attenuated in an ordinary medium, while the frequency and phase are preserved. If we reverse this absorption process in time, the light which is at first weak appears to progress in the opposite direction with increasing amplitude, but without any change in frequency or phase. This describes the behavior of light waves being amplified by a laser. How each individual atom contributes to the laser amplification will be described in Chap. 7.

5.5 Conditions for Laser Oscillation

If positive feedback is applied to an amplifier, the amplification is increased, although it may become unstable and oscillation may build up under certain

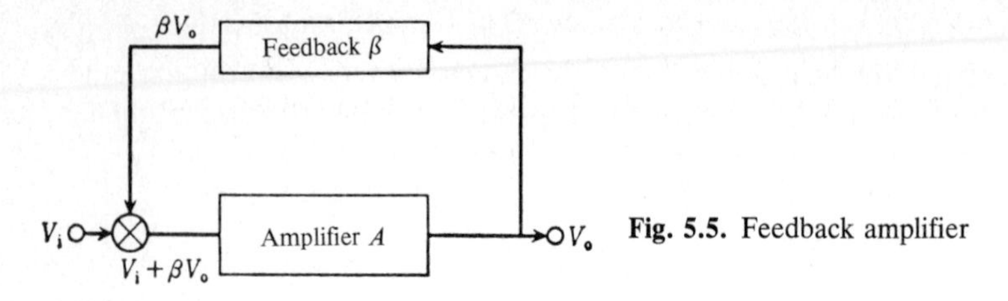

Fig. 5.5. Feedback amplifier

conditions. A feedback amplifier is shown in Fig. 5.5, where a fraction $\beta\,(< 1)$ of the output voltage V_o is fed back to be superposed on the input voltage V_i of the amplifier. When $\beta = 0$, no feedback is given and we have $V_o = A\,V_i$, where A is the voltage gain at the signal frequency. If $\beta \neq 0$, the effective input voltage becomes $V_i + \beta\,V_o$, which will be amplified to deliver

$$V_o = A\,(V_i + \beta\,V_o)\ .$$

Then, the voltage gain of the feedback amplifier is expressed as

$$\frac{V_o}{V_i} = \frac{A}{1 - \beta A}\ . \tag{5.22}$$

Since we are dealing with complex input and output voltages, both A and β are, in general, complex quantities involving their phase shifts. Accordingly, we have positive feedback when $|1 - \beta A| < 1$, and negative feedback when $|1 - \beta A| > 1$.

In order to make matters simple, suppose that there is phase shift neither in the amplifier nor in the feedback network, so that both A and β are real. Then we have positive feedback when $\beta A > 0$, and the voltage gain is greater than without feedback. If β or A is gradually increased to make $\beta A = 1$, the voltage gain will be infinite. Practically, in this case, even if the input is 0, small electric noises in the circuit elements and the amplifier are amplified to generate a large output voltage. The reason why the output voltage does not increase infinitely is because the energy that the amplifier can produce is limited; in other words, the amplification must always saturate.

Generally, both the voltage gain A and the feedback coefficient β are frequency dependent. Then the oscillation starts at a frequency at which βA becomes a maximum. As the oscillation builds up to a large amplitude, the nonlinear characteristic of the amplifier will reduce the voltage gain and more or less modify the frequency characteristics. Thus, the amplitude of oscillation, which initially increases exponentially, tends to reduce its rate of increase, accompanied by a shift in frequency. Finally, when the amplitude and the frequency reach the values that satisfy $\beta A = 1$, a steady-state oscillation with a constant amplitude will hold. It does not always follow, however, that the final state is of constant amplitude. In some cases, the value of βA

never reaches 1, and the amplitude increases and decreases alternately, together with some increase and decrease in frequency. Such a state of oscillation is often called relaxation oscillation. Again, under certain conditions, both the amplitude and the frequency change spasmodically, giving rise to a state known as chaos. The three types of oscillations described above are known to occur also with lasers[3].

We shall first treat the steady-state oscillation of the laser. We assume that there is a laser medium with population inversion between the mirrors M_1 and M_2, as shown in Fig. 5.1. Let R_1 and R_2 be the reflectance of the mirrors, l the length of the laser medium, and G the gain constant. Then the round-trip gain, while the light wave passes through the laser medium twice, is

$$A^2 = e^{2Gl} , \tag{5.23}$$

where A is the magnitude of the round-trip amplification. Since the round-trip feedback of the amplified output is $\sqrt{R_1 R_2}$, we have

$$\beta = \sqrt{R_1 R_2}\, e^{i\theta} . \tag{5.24}$$

The additional factor $\exp(i\theta)$ is necessary because the phase of the fed-back wave may be considerably different from that of the initial wave. The light wave has such a high frequency and short wavelength that this phase difference may change remarkably, even for a relatively small change in frequency. Therefore, the laser oscillates at a frequency for which θ is nearly 0, assuming no phase shift in amplification[4].

Thus, putting $\theta = 0$ and substituting the condition $\beta A = 1$ into the two equations above, we obtain

$$Gl + \tfrac{1}{2} \ln R_1 R_2 = 0 . \tag{5.25}$$

In an actual laser medium there are, besides atoms in the two levels relevant for laser action, atoms in other levels and some impurity atoms. Furthermore, there exists a host crystal in a solid-state laser, and ions and electrons in a gas laser, all of which absorb laser light to some extent. Such power losses, other than that due to the mirrors, can be expressed by a phenomenological round-trip power transmission coefficient $K (< 1)$. Then we may use the effective loss constant,

$$L_{\text{eff}} = -\frac{1}{2l} \ln K .$$

3 Undamped relaxation oscillation in a ruby laser was ascribed to nonlinear absorption by the author early in 1963 [5.3]. Damped relaxation oscillation in lasers will be treated in Sect. 6.4. Chaotic laser oscillation has recently been demonstrated [5.4].

4 This is not true in a high-gain laser medium.

With this additional loss, the condition (5.25) for laser oscillation is modified to become

$$(G - L_{\text{eff}})\,l + \tfrac{1}{2}\ln R_1 R_2 = 0\ , \quad \text{or} \tag{5.26}$$

$$G = \frac{1}{2l}\left|\ln R_1 R_2\right| + L_{\text{eff}}\ . \tag{5.27}$$

Since $R_1 < 1$ and $R_2 < 1$, we have $\ln R_1 R_2 < 0$ in these equations. Hence, the smaller the value of $|\ln R_1 R_2|$, the smaller the gain needed to achieve laser oscillation. Therefore we use mirror reflectances of close to 1. With the approximation $R_1 R_2 \approx 1$, (5.27) can be rewritten as

$$G = \frac{1 - R_1 R_2}{2l} + L_{\text{eff}}\ . \tag{5.28}$$

The left-hand side is the gain coefficient and the right-hand side is the loss coefficient so that the equation expresses a balance between gain and loss.

When the gain is smaller than the loss, the optical field, if any, decays exponentially. On the other hand, if the gain is greater than the loss, the oscillation builds up exponentially from any small optical electric field, such as the thermal radiation. The rate of increase in this case is given by the difference between gain and loss. As the optical electric field becomes strong enough, the number of atoms in the upper level is reduced and the gain of the laser medium becomes smaller, while the loss does not change appreciably. As the light intensity increases, therefore, the difference of gain over loss decreases, as shown in Fig. 5.6. Consequently, the rate of increase becomes smaller and finally reaches the point at which the gain equals the loss (point P in Fig. 5.6). Thus (5.27 or 28) gives the threshold value of the gain constant for laser oscillation, and at the same time gives the value at steady-state oscillation.

From (5.28, 17) the threshold value of population inversion ΔN_{th} or the population inversion under steady-state oscillation is given by

$$\Delta N_{\text{th}} = \frac{c}{\hbar\omega B(\omega)}\left(\frac{1 - R_1 R_2}{2l} + L_{\text{eff}}\right)\ . \tag{5.29}$$

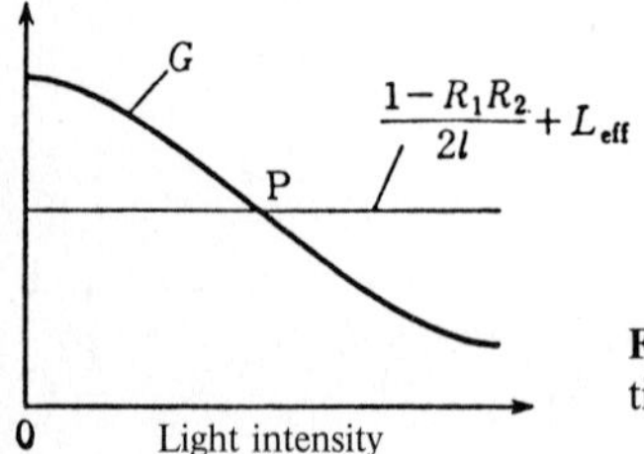

Fig. 5.6. Saturated gain G and the optical loss (5.28) as functions of the light intensity

It may be rewritten, using (5.21), as

$$\left(N_2 - \frac{g_2}{g_1} N_1\right)_{\text{th}} = \frac{3\,\varepsilon_0 c\hbar}{\pi\omega\,|\mu_{21}|^2 g(\omega)}\left(\frac{1 - R_1 R_2}{2l} + L_{\text{eff}}\right). \tag{5.30}$$

For the sake of simplicity, now the levels are assumed to be non-degenerate, and $B(\omega)$ and $g(\omega)$ are assumed to have a Lorentzian profile, with $g(\omega_0) = (\pi\Delta\omega)^{-1}$. Then (5.29) or (5.30) at $\omega = \omega_0$ becomes

$$\Delta N_{\text{th}} = \frac{3\,\varepsilon_0 c\hbar}{|\mu_{21}|^2} \cdot \frac{\Delta\omega}{\omega}\left(\frac{1 - R_1 R_2}{2l} + L_{\text{eff}}\right). \tag{5.31}$$

The first factor on the right-hand side shows that the greater the transition dipole moment the lower the threshold. The threshold is also lowered as the linewidth and the loss in the optical resonator are reduced, as seen from the second and third factors, respectively.

In the above treatment we have considered the amplification and decay of light traveling back and forth between the mirrors. We can also treat laser oscillation by considering the modes of the electromagnetic waves in the optical resonator.

The two mirrors of the laser shown in Fig. 5.1 act as a Fabry-Perot resonator (described in Sect. 3.4) and confine light with a wave vector perpendicular to the mirror surfaces and a frequency equal to one of the proper frequencies. Let W be the light energy stored in the resonator and P_{L} the energy loss per unit time, i.e., the power loss at the mirrors and in the laser medium. Then the quality factor of the resonator is defined as

$$Q_{\text{c}} = \frac{\omega W}{P_{\text{L}}} \tag{5.32}$$

where ω is the circular frequency of the light. Again, if κ denotes the decay constant of the amplitude in free natural decay in the resonator, we have

$$2\kappa = -\frac{1}{W}\frac{dW}{dt} = \frac{\omega}{Q_{\text{c}}}. \tag{5.33}$$

For simplicity, let us assume that the distribution of light intensity is uniform and that the laser medium is uniformly filled in the resonator ($l = L$). Then the power loss is expressed as

$$P_{\text{L}} = \frac{\omega U L}{Q_{\text{c}}}, \tag{5.34}$$

where U is the average energy density per unit length of the resonance mode. On the other hand, the light power P_{G} supplied to the resonance mode by stimulated emission from the laser medium, which has a population inversion ΔN, is given from (4.40):

$$P_G = \Delta N \hbar \omega B(\omega) UL . \tag{5.35}$$

Thus, if the population inversion for a gain P_G equal to the power loss P_L is denoted by ΔN_{th}, we obtain

$$\frac{1}{Q_c} = \Delta N_{th} \hbar B(\omega) . \tag{5.36}$$

The reflection loss of energy in an optical resonator composed of two mirrors separated by L with reflectances R_1 and R_2 is equal to $(1 - R_1 R_2) UL$, in a time $2L/c$. Therefore, neglecting power loss from other sources, we obtain the power loss P_L:

$$P_L = \frac{c(1 - R_1 R_2)}{2} U .$$

From this and (5.34), the quality factor of the resonator can be expressed in the form

$$Q_c = \frac{2\omega L}{c(1 - R_1 R_2)} . \tag{5.37}$$

The threshold value of population inversion or the population inversion for the steady state can be found from (5.36, 37) to be

$$\Delta N_{th} = \frac{c(1 - R_1 R_2)}{2\hbar\omega B(\omega) L} . \tag{5.38}$$

Since we have neglected here losses other than from reflection, this result is consistent with (5.29), which was derived by considering light traveling back and forth between the mirrors.

5.6 Frequency of Laser Oscillation

Let us now consider how the frequency is determined in a laser where a plane wave is traveling back and forth between the two mirrors. First, the complex representation of a plane wave progressing in the $+z$ direction from the mirror surface at $z = 0$ is written as

$$E(z, t) = E_0 \mathrm{e}^{\mathrm{i}(\omega t - kz)} \tag{5.39}$$

where k is a complex quantity. Denoting the real part of k by k_0, and since twice the imaginary part of k is the gain constant G, we have

$$k = k_0 + \mathrm{i}\frac{G}{2} = \frac{\omega}{c}\sqrt{1+\chi(\omega)}\,, \tag{5.40}$$

where $\chi(\omega)$ is the complex susceptibility, which is related to ε by $\varepsilon = \varepsilon_0[1+\chi(\omega)]$, as stated in Sect 4.6. If the magnitude of $\chi(\omega)$ is small, we can approximate the wavenumber k_0 and the gain constant G in the form

$$k_0 = \frac{\omega}{c}\left[1 + \frac{1}{2}\chi'(\omega)\right] \quad \text{and} \tag{5.41}$$

$$G = -\frac{\omega}{c}\chi''(\omega)\,. \tag{5.42}$$

Since $\chi'' < 0$ for population inversion, we have $G > 0$.

By substituting (5.40) into (5.39), we obtain

$$E(L,t) = E_0\,\mathrm{e}^{(1/2)\,GL}\,\mathrm{e}^{\mathrm{i}\,(\omega t - k_0 L)} \tag{5.43}$$

on the mirror surface at $z = L$. If the amplitude reflection coefficients of the mirrors at $z = 0$ and L are r_1 and r_2, respectively, the wave in (5.43) is reflected at $z = L$ with its amplitude multiplied by r_2 and it progresses in the $-z$ direction. When it is reflected next by the mirror at $z = 0$, it must be further multiplied by r_1 and expressed as

$$E(0,t) = r_1 r_2 E_0\,\mathrm{e}^{GL}\,\mathrm{e}^{\mathrm{i}\,(\omega t - 2\,k_0 L)}\,,$$

progressing in the $+z$ direction. Now, when the laser oscillation is steady, the above expression must be equal to the initial value $E_0 \exp(\mathrm{i}\omega t)$ at $z = 0$ which is assumed in (5.39). Therefore, the condition for steady-state oscillation is given by

$$r_1 r_2\,\mathrm{e}^{GL}\,\mathrm{e}^{-2\,\mathrm{i}k_0 L} = 1\,.$$

By using $r_1 r_2 = \sqrt{R_1 R_2}\exp(\mathrm{i}\theta)$ which is the same as β of (5.24), the absolute value of the above equation becomes

$$\sqrt{R_1 R_2}\,\mathrm{e}^{GL} = 1\,, \tag{5.44}$$

while the phase angle is rewritten as

$$2\,k_0 L = 2\,n\pi + \theta\,, \tag{5.45}$$

where n is an integer. The logarithm of (5.44) is exactly the same as (5.25) obtained from the energy balance. Substitution of (5.42) gives

$$\frac{\omega}{c}\chi''(\omega)L = \frac{1}{2}\ln R_1 R_2 \tag{5.46}$$

where both sides are negative. Moreover, if $R_1 R_2$ is close to 1, we may use the approximation $\ln R_1 R_2 = R_1 R_2 - 1$ and rewrite the above equation for the condition of laser oscillation as

$$-\chi''(\omega) = \frac{1}{Q_c}, \tag{5.47}$$

where Q_c, the quality factor of the optical resonator, given by (5.37) is used.

It is (5.45) that determines the frequency of oscillation. By using (5.41) it becomes

$$\frac{\omega}{c}[2+\chi'(\omega)]L = 2n\pi + \theta\,. \tag{5.48}$$

From (4.52, 53) or from (4.58, 59) it can be seen that the relation

$$\chi'(\omega) = -\chi''(\omega)\frac{\omega-\omega_0}{\gamma} \tag{5.49}$$

holds for the Lorentzian lineshape. It is rewritten for the steady-state oscillation condition (5.47) as

$$\chi'(\omega) = \frac{\omega-\omega_0}{\gamma Q_c}\,.$$

Therefore, by substituting this into (5.48), the frequency[5] ω of the steady state is given by

$$\frac{\omega}{c}\left(2+\frac{\omega-\omega_0}{\gamma Q_c}\right)L = 2n\pi + \theta\,. \tag{5.50}$$

Taking the phase shift on reflection into consideration, the resonant frequency ω_c is expressed as

$$\omega_c = \frac{c}{L}\left(n\pi + \frac{\theta}{2}\right) \tag{5.51}$$

---

5 The circular frequency or the angular frequency is written simply as the frequency.

by putting $\chi' = 0$ in (5.48). Substituting (5.51) into the right-hand side of (5.50), and using (5.33) for ω/Q_c on the left-hand side, we obtain

$$\frac{\omega - \omega_c}{\kappa} + \frac{\omega - \omega_0}{\gamma} = 0 \, . \tag{5.52}$$

Thus, the frequency of the laser is found to be

$$\omega = \frac{\kappa\omega_0 + \gamma\omega_c}{\kappa + \gamma} \, . \tag{5.53}$$

Since κ is the decay constant of the resonator, it is equal to the halfwidth (in angular frequency) at half maximum of the resonance curve. Again, γ is the decay constant of the oscillating dipole, namely, the halfwidth at half maximum of the spectral line. If we define the Q value of the spectral line as $Q_0 = \omega_0/2\gamma$, which expresses the sharpness, we may rewrite (5.53) in the form

$$\omega = \frac{Q_0\omega_0 + Q_c\omega_c}{Q_0 + Q_c} \, , \tag{5.54}$$

using the approximation $Q_c = \omega/2\kappa \approx \omega_0/2\kappa$.

It is seen from these equations that the frequency of the laser lies between the resonant frequency of the optical resonator and the frequency of the laser transition; it is nearer to the frequency of the one having a higher Q. In an ordinary laser, the Q of the resonator is greater than that of the spectral line, i.e. $Q_c > Q_0$; therefore, the laser frequency is, to a first approximation, equal to the resonant frequency of the resonator. However, it is somewhat shifted towards the frequency of the spectral line. This is called the frequency pulling effect.

Most lasers oscillate not only in a single mode but in many modes. This is because the spectral linewidth γ is wider than the mode separation $c\pi/L$ given by (5.51), resulting in oscillations in a number of modes on which the small-signal gain exceeds the loss. In this chapter we have considered only the longitudinal modes of the Fabry-Perot resonator. It should be mentioned that laser oscillations in transverse modes of higher orders are also found in most practical lasers.

Problems

5.1 Find the relation between the rate of excitation Γ and the power density of pumping P in an optically pumped three-level system as shown in Fig. 5.2.

Assume that the relaxation rates $\gamma_{21}(\gg\gamma_{12})$, $\gamma_{31}(\gg\gamma_{13})$, and $\gamma_{32}(\gg\gamma_{23})$ are independent of P.

Hint: $\Gamma = (N_1 - N_3)\,BP/c$.
Answer: $\Gamma = \{[(\gamma_3 + bP)^2 + 4\gamma_3 abP]^{1/2} - \gamma_3 - BP\}/2a$, where $\gamma_3 = \gamma_{31} + \gamma_{32}$, $a = (\gamma_{21} + \gamma_{32})/\gamma_{21}$, and $b = BN/c$.

5.2 Find the steady-state population inversion in a three-level system as shown in Fig. 5.2 in the case when the pumping is so strong as to completely saturate the transition between level 1 and level 3, while γ_{12}, γ_{13}, and γ_{23} are not negligible.

Answer: $N_2 - N_1 = N_2 - N_3 = (\gamma_{12} + \gamma_{32} - \gamma_{21} - \gamma_{23})/[\gamma_{12} + \gamma_{32} + 2(\gamma_{21} + \gamma_{23})]$.

5.3 A gas of three-level atoms is initially in equilibrium at a high temperature T_a. Suppose that the temperature of the gas is suddenly reduced to a lower temperature T_b so that atomic populations in the excited levels change with time, reaching eventually their equilibrium values at T_b. Show that the condition for transient inversion of populations between level 2 and level 3, when $\gamma_{32} = 0$, is given by

$$(K-1)^{K-1}(a_3 - b_3)^K > K^K (b_2 - b_3)^{K-1}(a_2 - b_2)$$

where

$$a_2 = \exp[-(W_2 - W_1)/k_B T_a],\ a_3 = \exp[-(W_3 - W_1)/k_B T_a],$$
$$b_2 = \exp[-(W_2 - W_1)/k_B T_b],\ b_3 = \exp[-(W_3 - W_1)/k_B T_b], \quad \text{and}$$
$$K = \gamma_{21}/\gamma_{31} > 1\ .$$

Hint: $N_3/N_1 = (a_3 - b_3)\exp(-\gamma_{31}t) + b_3$,
$N_2/N_1 = (a_2 - b_2)\exp(-K\gamma_{31}t) + b_2$,

and maximum inversion appears at time t when

$$\exp[(K-1)\gamma_{31}t] = K(a_2 - b_2)/(a_3 - b_3)\ .$$

5.4 Find the expression for the gain constant at the center of the Lorentzian profile in terms of the Einstein A coefficient, the population inversion, and the wavelength.

Answer: $G = \left(N_2 - \dfrac{g_2}{g_1}N_1\right)\dfrac{A\lambda^2}{8\pi^2\Delta\nu}$, where $\Delta\nu$ is the halfwidth at half maximum of the Lorentzian profile.

5.5 Consider that both ends of a ruby laser rod of 5 cm length are coated to have a reflectance of $R = 0.9$. What is the minimum fraction of excited Cr ions achieving the threshold condition of oscillation? Assume that the concentration of Cr ions is $N = 1 \times 10^{19}\ \text{cm}^{-3}$, the induced-emission cross-section is $\sigma = 2 \times 10^{-20}\ \text{cm}^2$, and the effective loss constant of the rod is $0.011\ \text{cm}^{-1}$.

Answer: $(N_2/N)_{\text{min}} = 58\%$, since $\Delta N_{\text{th}} = 1.6 \times 10^{18}\ \text{cm}^{-3}$.

5.6 Find the threshold condition of a laser with a gain medium of length L_g and an absorber of length L_a in a resonator of length $L_a + L_g$. Express the threshold population inversion in terms of the induced-emission cross-section of the gain medium σ, the amplitude absorption constant of the absorber α, and the mirror reflectance R.

Answer: $\Delta N_{\text{th}} = (2\alpha L_a + |\ln R|)/\sigma L_g$.

5.7 According to (5.31), the larger the transition dipole moment the lower the threshold of population inversion. On the other hand, in practice, laser oscillation is easier when using a fairly weak emission line of the medium than when using a strong line corresponding to a transition with a large dipole moment. Why is it hard to achieve laser oscillation based on the transition with a large dipole moment?

Answer: When the transition dipole moment is larger, the lifetime of the relevant level is shorter so that the atomic population in the excited state will be less. Thus it will be harder to achieve population inversion. Moreover, the broader linewidth, because of the larger collisional cross-section and the shorter lifetime, tends to raise the threshold value of population inversion as seen from (5.31).

5.8 Evaluate the coefficient of frequency pulling by the resonator from the frequency of the laser transition as defined by $S = (\omega - \omega_0)/(\omega_c - \omega_0)$. What is the value of S in a laser filled with a medium of 50 cm length, having a Lorentzian gain profile of 0.5 GHz halfwidth at half maximum, terminated by mirrors of 90% reflectance?

Answer: $S = 0.991$.

6. Output Characteristics of the Laser

The characteristics of the laser are determined by the interaction between the light and the atoms in the resonator under the influence of pumping and relaxation. There are different theories of the laser in treating these factors. They are classified into the rate-equation theory, the semiclassical theory, and the quantum-mechanical theory. In the quantum-mechanical theory, both light and atoms are treated quantum-mechanically, while in the semiclassical theory the light field is treated as classical electromagnetic waves according to Maxwell's equations, and the atoms are treated quantum-mechanically. The rate equations describe temporal changes of atomic populations and light energy or the number of photons, neglecting the phase of the light and the atomic polarization. Therefore, the rate equations cannot be applied to subjects such as the frequency of the laser or the characteristics of multimode oscillation.

6.1 Rate Equations of the Laser

In Chap. 5 we discussed the conditions leading to population inversion in three-level and four-level systems. The population inversion was calculated for the case without stimulated emission. Now, when the laser is operating, the population inversion is reduced by the effect of stimulated emission. Moreover, actual atoms are multi-level systems and the laser action is affected by the additional levels. Since it is the purpose of this chapter to investigate the power output of the laser, the effect of levels other than those of the laser transition, and of pumping, will only be introduced as parameters in the relaxation rate and the excitation rate. We will not discuss the process of excitation, but discuss the effect of stimulated emission caused by the laser power.

We consider that the laser transition takes place from level 2 to level 1, whose relaxation rates (sometimes called relaxation constants) are, respectively, γ_2 and γ_1. Atoms are preferentially pumped to the upper level, but some atoms are excited to the lower level; the excitation rates (the number of atoms per unit volume excited in a unit time) of the upper and lower levels are then denoted by Φ_2 and Φ_1, respectively. Although these relaxation rates and excitation rates vary considerably with material, temperature, and strength of pumping, etc., they are relatively independent of the light intensity so that they can be assumed to be constant.

Suppose that atoms in a laser medium are in a laser field of energy density W and frequency ω. The number of atoms per unit volume in the upper and lower levels are, respectively, N_2 and N_1, hereafter called the populations. The probability of stimulated emission per unit time is $N_2 B(\omega) W$ and the probability of absorption is $N_1 B(\omega) W$, where $B(\omega)$ is defined by (4.39) in Sect 4.5. Thus, the rates of change of population in the upper and lower levels are, respectively,

$$\frac{dN_2}{dt} = \Phi_2 - \gamma_2 N_2 - (N_2 - N_1) B(\omega) W , \tag{6.1}$$

$$\frac{dN_1}{dt} = \Phi_1 - \gamma_1 N_1 + (N_2 - N_1) B(\omega) W . \tag{6.2}$$

Denoting the amplitude decay constant of the resonator by κ, we can write the rate of change of the optical energy density in the resonator as

$$\frac{dW}{dt} = -2\kappa W + \hbar\omega (N_2 - N_1) B(\omega) W . \tag{6.3}$$

The second term on the right-hand side is the increase of energy caused by stimulated emission; the effect of spontaneous emission is neglected as being small compared to stimulated emission. Equations (6.1–3) are the fundamental rate equations of the laser.

Let us use the superscript (0) for the populations in the case when the laser power is absent, $W = 0$, while the laser medium is being pumped. Then, from (6.1, 2), we obtain

$$N_2^{(0)} = \frac{\Phi_2}{\gamma_2} , \quad N_1^{(0)} = \frac{\Phi_1}{\gamma_1} , \tag{6.4}$$

for the steady state. In thermal equilibrium the excitation is due to themal excitation only, so that the ratios of thermal excitation to relaxation rate become

$$\left(\frac{\Phi_2}{\gamma_2}\right)_T \propto \exp\left(-\frac{W_2}{k_B T}\right) , \quad \left(\frac{\Phi_1}{\gamma_1}\right)_T \propto \exp\left(-\frac{W_1}{k_B T}\right) ,$$

as in (5.1). Then the ratio of the atomic populations is

$$\left(\frac{N_2}{N_1}\right)_T = \exp\left(-\frac{W_2 - W_1}{k_B T}\right) ,$$

which follows the Boltzmann distribution. The population inversion

$$\Delta N^{(0)} = N_2^{(0)} - N_1^{(0)} \tag{6.5}$$

in the laser can be made large by having a large Φ_2 and a small γ_2 using various methods mentioned before.

The condition for laser oscillation, or attaining a threshold value of the population inversion, is given by the value of $N_2 - N_1$ which makes the right-hand side of (6.3) equal to zero. It is

$$\Delta N_{\mathrm{th}} = \frac{2\kappa}{\hbar\omega B(\omega)} . \tag{6.6}$$

This is exactly the same result as that obtained from the treatment in Sect 5.5, if (5.33) is substituted into (5.36).

6.2 Steady-State Output

The rate equations (6.1–3) are equal to zero for the steady state. Taking the difference between (6.1)/γ_2 and (6.2)/γ_1, and using (6.4), we obtain

$$\Delta N^{(0)} - \Delta N - 2\tau \Delta N B(\omega) W = 0 , \quad \text{where} \tag{6.7}$$

$$\Delta N = N_2 - N_1 , \qquad \tau = \frac{1}{2}\left(\frac{1}{\gamma_2} + \frac{1}{\gamma_1}\right) . \tag{6.8}$$

Here, τ represents the effective relaxation time of the two-level system. From (6.7) we obtain

$$\Delta N = \frac{\Delta N^{(0)}}{1 + 2\tau B(\omega) W} . \tag{6.9}$$

This shows that the population inversion ΔN of the laser medium decreases with the optical energy density W, as shown in Fig. 6.1. As can be seen from the right-hand side of (6.3), the rate of increase of the optical energy is reduced as ΔN becomes smaller; when the absolute values of the two terms in the right-hand side become equal, we have a steady state. Then we obtain

$$\Delta N = \frac{2\kappa}{\hbar\omega B(\omega)} \tag{6.10}$$

namely, $\Delta N = \Delta N_{\mathrm{th}}$.

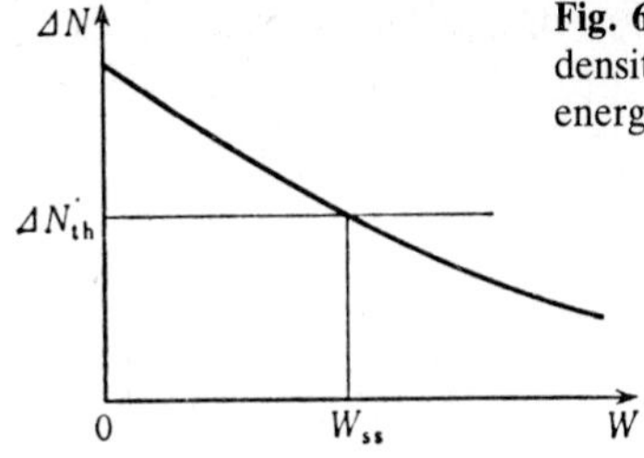

Fig. 6.1. Decrease of the population inversion with the energy density W. It becomes equal to the threshold value ΔN_{th} at the energy density W_{ss} of the steady-state laser

Since the energy density W_{ss} in the steady state is the value of W satisfying both (6.9) and (6.10), by putting $\Delta N = \Delta N_{\text{th}}$ we obtain

$$W_{\text{ss}} = \frac{\hbar\omega}{4\kappa\tau}(\Delta N^{(0)} - \Delta N_{\text{th}}) \quad \text{or} \tag{6.11 a}$$

$$W_{\text{ss}} = \frac{1}{2\tau B(\omega)}\left(\frac{\Delta N^{(0)}}{\Delta N_{\text{th}}} - 1\right). \tag{6.11 b}$$

As the intensity of excitation is gradually increased, the laser output appears when the population inversion exceeds the value ΔN_{th}. As shown in Fig. 6.2, the output power increases linearly in proportion to $\Delta N^{(0)} - \Delta N_{\text{th}}$, the excess value above the threshold. This relation is known to hold approximately for any laser. However, the assumptions made in the calculation above do not hold rigorously for each actual laser. For example, when the excitation is strong, the temperature of the medium rises so that the relaxation constant may change and the power characteristic may deviate from a straight line. The power characteristic curve of some lasers rises above the straight line as ΔN is increased, while that of others falls below the straight line. Moreover, if we plot the power characteristic against the excitation intensity, e.g., the pumping current, instead of the population inversion, the deviation from a straight line is often greater.

Equation (6.11) gives the energy density in the resonator. Since the optical power emitted by stimulated emission from the laser medium is $P_{\text{ss}} = 2\kappa W_{\text{ss}}$ per unit volume, we obtain from (6.11 a),

$$P_{\text{ss}} = \frac{\hbar\omega}{2\tau}(\Delta N^{(0)} - \Delta N_{\text{th}}). \tag{6.12}$$

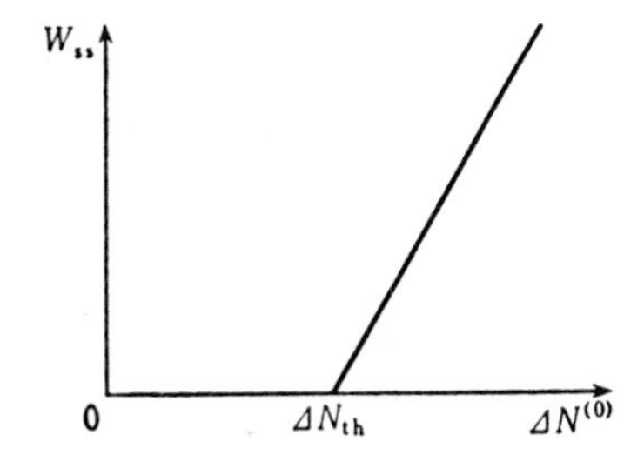

Fig. 6.2. Power output of the steady-state laser as a function of population inversion

Since τ in the denominator of (6.12) is the relaxation time of the atom as defined by (6.8), $\hbar\omega/\tau$ is the optical power emitted by a single atom. Now, when an atom makes a transition from the upper level to the lower level, N_2 decreases by 1 and N_1 increases by 1, the population inversion thereby decreasing by 2. Thus, by noting that the effective number of atoms that can emit light is $\Delta N/2$, (6.12) can be easily understood.

The equation above is for the power inside the resonator and not the power available outside. One of the mirrors in an ordinary laser is partially transparent so that an output can be extracted, with a consequent loss of power in the resonator and a rise in the threshold value. Thus, if the transmittance of the mirror is too large, the output power will drop. There is an optimum value of the transmittance for maximum output.

Denoting the transmittance of one of the mirrors of the resonator of length L by T, the steady-state output per unit volume of the laser medium can be expressed as

$$P_{\text{out}} = 2\,\kappa_T W_{\text{ss}}\,, \qquad \kappa_T = \frac{c}{4L} T\,, \tag{6.13}$$

where κ_T is the amplitude decay constant due to the output coupling mentioned above, which may also be called the output coupling coefficient. If we denote the amplitude decay constant by κ_0 when $T = 0$, we may presume, in general, that

$$\kappa = \kappa_0 + \kappa_T\,, \tag{6.14}$$

so that, using (6.11 a), the output power per unit volume may be written

$$P_{\text{out}} = \frac{\hbar\omega\kappa_T}{2\,\kappa\tau} (\Delta N^{(0)} - \Delta N_{\text{th}})\,. \tag{6.15}$$

Since the threshold ΔN_{th} varies in proportion to $\kappa = \kappa_0 + \kappa_T$, as shown in (6.6), it takes a minimum value when $\kappa_T = 0$:

$$\Delta N_{\text{th, min}} = \frac{2\,\kappa_0}{\hbar\omega B(\omega)}\,. \tag{6.16}$$

Using the relative excitation $\mathcal{N}$, defined by the value of the population inversion relative to its minimum,

$$\mathcal{N} = \frac{\Delta N^{(0)}}{\Delta N_{\text{th, min}}}\,, \tag{6.17}$$

we may rewrite (6.15)

$$P_{\text{out}} = \frac{\kappa_T}{\tau B(\omega)} \left(\frac{\kappa_0 \mathcal{N}}{\kappa_0 + \kappa_T} - 1 \right). \tag{6.18}$$

Then, the output power as a function of the output coupling coefficient is obtained as shown in Fig. 6.3. The optimum coupling κ_{opt} for a maximum output power is the optimum value of κ_T, given by

$$\kappa_{opt} = (\sqrt{\mathcal{N}} - 1)\,\kappa_0 \,, \tag{6.19}$$

which is obtained by differentiating (6.18) with respect to κ_T and equating to zero. The output power available at this optimum coupling can then be expressed as

$$P_{opt} = \frac{\kappa_0}{\tau B(\omega)}\,(\sqrt{\mathcal{N}} - 1)^2 \,. \tag{6.20}$$

6.3 Oscillation Build-Up

The build-up of laser oscillation can be investigated using the rate equations (6.1–3). As mentioned in Sect. 5.6, the Q value of the optical resonator is greater than that of the spectral line in an ordinary laser, i.e., $\kappa \ll \gamma$. Thus, as long as the rate of increase of energy in the resonator due to stimulated emission is not too large, the change in optical energy W is slower in comparison with the change in population inversion ΔN. Therefore, when considering the temporal variation of W in (6.3), ΔN may be assumed to take a quasi-steady value so that the build-up can be easily calculated as follows. The case when the changes in both ΔN and W must be taken into consideration together will be treated in the next section.

Here, we shall assume that as W increases gradually during the build-up of oscillation, ΔN takes the steady-state value (6.9) determined by the instantaneous value of the light intensity. Substituting (6.9) into (6.3), we obtain

$$\frac{dW}{dt} = -2\kappa W + \frac{\hbar\omega \Delta N^{(0)} B(\omega)\, W}{1 + 2\tau B(\omega)\, W} \,. \tag{6.21}$$

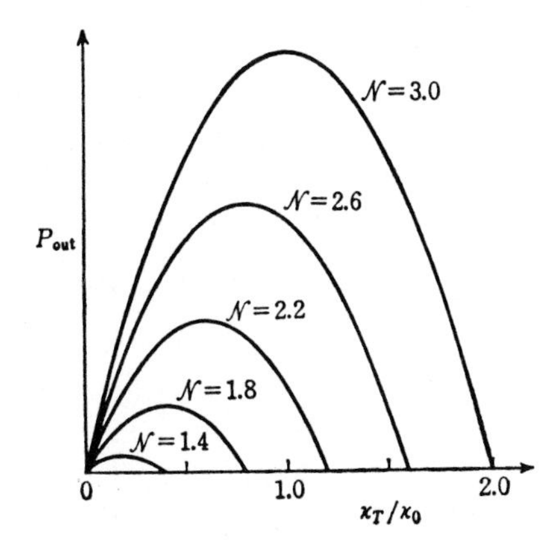

Fig. 6.3. Output power against the output coupling coefficient κ_T/κ_0. $\mathcal{N}$ is the relative intensity of excitation

To simplify the expressions, we put

$$\hbar\omega\Delta N^{(0)}B(\omega) = 2a\,, \quad 2\tau B(\omega) = b\,,$$

and use the approximation $(1 + bW)^{-1} \approx (1 - bW)$, assuming that the saturation is not too strong. Then the optical energy density in the steady state can be written as

$$W_{\mathrm{ss}} = \frac{a - \kappa}{ab}\,. \tag{6.22}$$

Equation (6.21) can be rewritten as

$$2\,dt = \frac{dW}{(a - \kappa - abW)\,W}\,. \tag{6.23}$$

Integration of (6.23) gives

$$2(a - \kappa)\,t = \ln \frac{W}{a - \kappa - abW} \cdot \frac{a - \kappa - abW_0}{W_0}\,,$$

where W_0 is the initial energy at $t = 0$, e.g., the thermal radiation energy. By using (6.22), the above equation can be rewritten as

$$\mathrm{e}^{2(a-\kappa)t} = \frac{W}{W_0} \cdot \frac{W_{\mathrm{ss}} - W_0}{W_{\mathrm{ss}} - W}\,.$$

Solving this for W, the temporal behavior of W can be expressed as

$$W(t) = \frac{W_{\mathrm{ss}} W_0\, \mathrm{e}^{2(a-\kappa)t}}{W_{\mathrm{ss}} + W_0\left(\mathrm{e}^{2(a-\kappa)t} - 1\right)}\,. \tag{6.24}$$

Since $W_{\mathrm{ss}} \gg W_0$ this may be rewritten to a good approximation as

$$W(t) = \frac{W_{\mathrm{ss}}}{1 + A\,\mathrm{e}^{-2(a-\kappa)t}}\,, \quad \text{where} \tag{6.25}$$

$$A = \frac{W_{\mathrm{ss}}}{W_0}\,.$$

Results of calculation for typical values of the parameters are shown in Figs. 6.4 and 5. The decay constant is $\kappa = c(1 - R_1 R_2)/4L$ from (5.33, 37), and takes a value $\kappa = 7.5 \times 10^6\,\mathrm{s}^{-1}$ when $L = 1$ m and $R_1 R_2 = 0.9$. We have taken the amplification coefficient a to be a little larger than κ so that

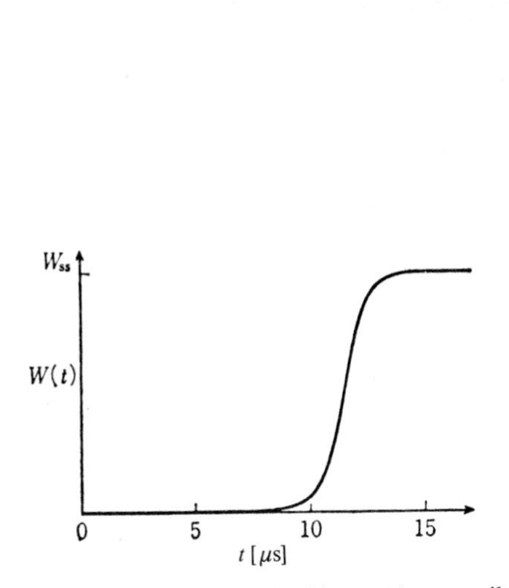

Fig. 6.4. Quasi-steady build-up of laser oscillation. ($a - \kappa = 1 \times 10^6$ s^{-1}, $A = 10^{10}$)

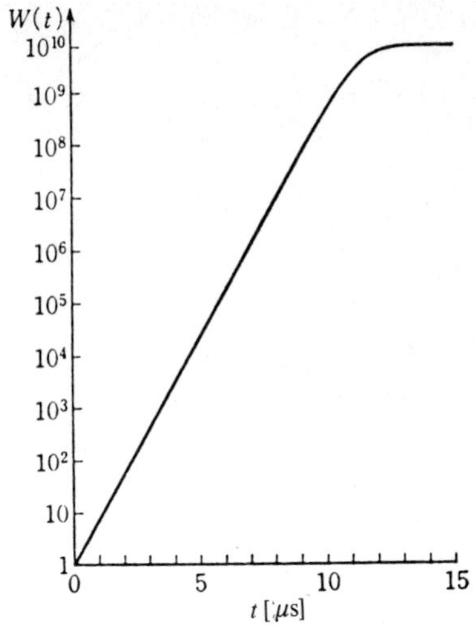

Fig. 6.5. Semi-logarithmic plot of the same build-up as in Fig. 6.4.

$a - \kappa = 1 \times 10^6 \mathrm{s}^{-1}$. The initial energy W_0 may be given by the thermal radiation at $t = 0$ or by the scattered pumping radiation in the laser mode. The value of W_0 is presumably about 1–$10^3 \hbar\omega$. However, it is immediately evident in (6.25) and in Fig. 6.5, where $W(t)$ is given on a logarithmic scale, that a large change in the value of W_0 is practically equivalent to a shift in time scale by a small amount. In this example we have put $A = W_{ss}/W_0 = 10^{10}$.

It is clearly seen in Fig. 6.5 that the laser power first increases exponentially, then saturation appears and finally it reaches a steady state. The greater part of the curve in Fig. 6.4 shows the variation after $t = 10$ μs when the saturation effect sets in. The above calculation of oscillation build-up holds approximately when $a - \kappa$ is very much smaller than τ^{-1}. A good example is a cw gas laser of low gain. On the other hand, for laser resonators with low Q and laser media of high gain, such as in solid-state lasers, the above approximation is not legitimate. Such cases are treated in the following sections.

6.4 Relaxation Oscillation

The build-up of a ruby laser excited by a flash lamp or of a semiconductor laser excited by a pulsed current often shows relaxation oscillations (Fig. 6.6) rather than the monotonic rise shown in Fig. 6.4. The characteristics of such relaxation oscillations or pulsed oscillations can be discussed by using rate equations. Since the calculation using the three equations (6.1–3) is troublesome, we shall assume that the pulse or the transient phenomenon appears

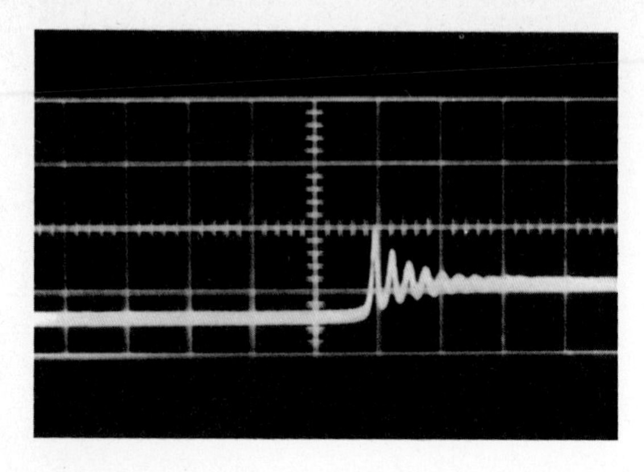

Fig. 6.6. Relaxation oscillation observed in a ruby laser. The time scale is 10 μs per division [6.1]

only for a relatively short time, and that practically no transitions (excitation or relaxation) other than that between the upper and lower laser levels occur during this time.

The sum $N_2 + N_1$ of the number of atoms in the upper and lower levels is constant in this case, so that the time variation of the population inversion $\Delta N = N_2 - N_1$ can be written as

$$\frac{d}{dt}\Delta N = \frac{\Delta N^{(0)} - \Delta N}{\tau} - 2\Delta NWB(\omega) , \quad (6.26)$$

and (6.3) becomes

$$\frac{d}{dt}W = -2\kappa W + \hbar\omega\Delta NWB(\omega) . \quad (6.27)$$

These equations constitute a set of nonlinear simultaneous differential equations, containing a term which has a product of the two variables W and ΔN.

In order to simplify the expressions, we transform the variables ΔN, W and t into non-dimensional quantities as follows:

$$x = \hbar\omega\tau B(\omega)\Delta N , \quad y = \tau B(\omega) W , \quad \theta = \frac{t}{\tau} .$$

Then (6.26 and 27) become

$$\frac{dx}{d\theta} = x_0 - x - 2xy , \quad (6.28)$$

$$\frac{dy}{d\theta} = -x_s y + xy , \quad (6.29)$$

where we have put

$$x_s = 2\kappa\tau , \quad x_0 = \hbar\omega\tau B(\omega)\Delta N^{(0)} .$$

It is evident that x_s is the threshold value of the population inversion (6.6), that is, the value of x for the steady state (6.10), and that, by putting (6.28 and 29) equal to zero, the steady-state energy is given by

$$y_s = \frac{1}{2}\left(\frac{x_0}{x_s} - 1\right), \tag{6.30}$$

which is equivalent to (6.11).

A result of the radiation build-up obtained by numerical calculation of the nonlinear simultaneous differential equations (6.28, 29) for a comparatively large value of the relative excitation $\mathcal{N} = x_0/x_s$ is given in Fig. 6.7. This appears to give a satisfactory explanation of the relaxation oscillation of an actual laser, shown in Fig. 6.6. After the initial large-amplitude relaxation oscillation has decayed and approached the steady state, the rate equations can be linearized and discussed analytically. It is then easier to inspect the effects of the parameters upon the period of relaxation oscillation and the decay constant.

When x and y are, respectively, close to x_s and y_s, we put

$$x = x_s(1 + p), \quad y = y_s(1 + q), \tag{6.31}$$

and neglect the term pq, since p and q are small compared to 1. Then (6.28, 29) are reduced to

$$\frac{dp}{d\theta} = -\mathcal{N}p - (\mathcal{N} - 1)q, \quad \frac{dq}{d\theta} = x_s p .$$

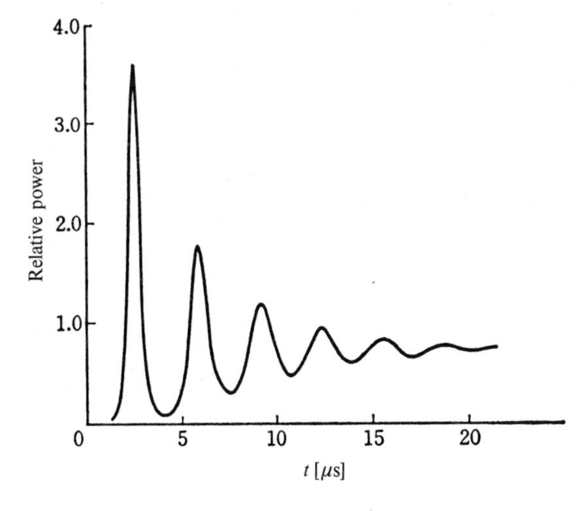

Fig. 6.7. An example of numerically calculated relaxation oscillation

Eliminating p from these two equations, we obtain

$$\frac{d^2q}{d\theta^2} + \mathcal{N}\frac{dq}{d\theta} + x_s(\mathcal{N}-1)q = 0 . \quad (6.32)$$

This is the well-known differential equation of a damped oscillation. Its solution as a function of $t = \theta\tau$ is

$$q = A\,\mathrm{e}^{-\gamma t}\cos\omega_m(t-t_0) , \quad (6.33)$$

where A and t_0 are initial values, and

$$\omega_m = \frac{1}{2\tau}\sqrt{4x_s(\mathcal{N}-1)-\mathcal{N}^2} \simeq \frac{1}{\tau}\sqrt{x_s(\mathcal{N}-1)} ,$$

$$\gamma = \frac{\mathcal{N}}{2\tau} .$$

It is seen from this result that the stronger the excitation the shorter the period of relaxation oscillation and the larger the decay constant. The above approximation for ω_m is for the case when $\kappa\tau \gg 1$ so that $x_s \gg 1$.

6.5 Q-Switching

When the lifetime of the upper level of the laser transition is relatively long, a short laser pulse of high instantaneous power can be obtained by a method that is known as Q-switching. In this mode of operation, the laser is initially pumped continuously while the Q of the laser resonator is kept low. When the population inversion has sufficiently increased, the Q value of the resonator is suddenly switched to a high value, with a result that the energy accumulated in the upper level is released as a laser output in a very short time. Since the peak power of the laser pulse obtained in such a way is considerably higher than that obtainable from ordinary pulsed oscillation, the output of the Q-switched laser is called a giant pulse.

The behavior of the Q-switched laser can also be investigated by the rate equations (6.28, 29). For an ordinary oscillation we have $y \lesssim y_s$, and it can be seen from (6.30) that y is of the order of 1. In a Q-switch operation, on the contrary, the term $x_0 - x$ in (6.28) can be neglected compared to $2xy$, since $y \gg 1$ except at the very beginning of the pulse build up. Thus we obtain

$$\frac{dx}{d\theta} = -2xy , \quad (6.34\,\mathrm{a})$$

$$\frac{dy}{d\theta} = -x_s y + xy . \quad (6.34\,\mathrm{b})$$

We can use these equations to calculate numerically the temporal changes of the normalized pulse power and the population inversion, as shown in Fig. 6.8, for example.

It can be seen in this figure that, as the population inversion takes a value of 10 times the threshold value at the beginning, the pulse builds up rapidly, and the population inversion decreases as the pulse grows, reaching a maximum, after which the pulse decays, since the gain of the laser medium is now reduced to practically nil. In the example of Fig. 6.8 the peak power of the pulse reaches a value of 7450 times the steady-state power.

Although such a pulse shape can only be obtained by a numerical calculation, an analytical relation between x and y can be obtained, from which the peak power and the pulse energy may be calculated as shown below. Rewriting (6.34 a) as

$$\frac{dx}{x} = -2y\,d\theta$$

and integrating it with the initial value $x = x_0$ at $\theta = 0$, we obtain

$$\ln\frac{x}{x_0} = -2\int_0^\theta y\,d\theta\,. \tag{6.35}$$

Again, using (6.34 a) to rewrite (6.34 b) as

$$dy = -x_s y\,d\theta - \tfrac{1}{2}\,dx\,,$$

integrating it, and substituting into (6.35), we obtain

$$y = \frac{x_0 - x}{2} - \frac{x_s}{2}\ln\frac{x_0}{x}\,. \tag{6.36}$$

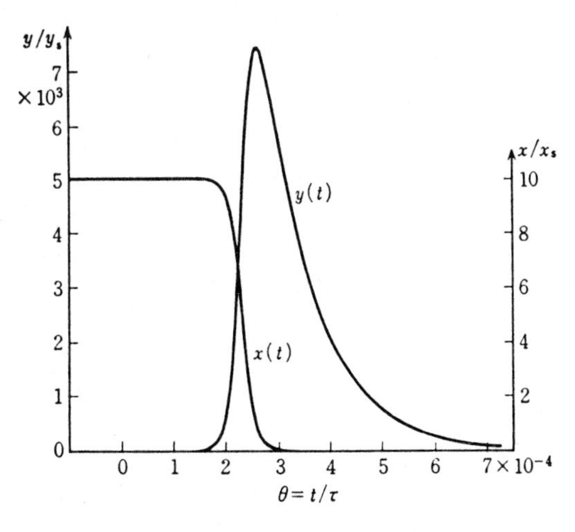

Fig. 6.8. The normalized output power $y(t)$ and the population inversion $x(t)$ of a Q-switched laser. Here, $x_s = 10^4$, $x_0 = 10x_s = 10^5$ are assumed, and the initial value of the normalized light intensity is taken to be $y(0) = 10^{-5}\,y_s$. From (6.30) we have $y_s = 4.5$; therefore $y(0) = 4.5 \times 10^{-5}$. (Calculated by Yoshio Cho)

Now this relation between x and y is graphically shown in Fig. 6.9, where the parameter x_0 is the normalized population inversion at $t = \theta = 0$.

The normalized power y is zero at the beginning when $x = x_0$. As the power y increases with time, the population inversion x decreases correspondingly. When x has decreased to x_s, the gain just equals the loss, and there will be no further increase in the power y, which gradually decreases from the maximum to become eventually zero. The final population inversion $\Delta N^{(f)}$ is given by

$$\Delta N^{(f)} = \frac{x_f}{\hbar\omega\tau B(\omega)} \tag{6.37}$$

where x_f is the value of x satisfying

$$x_0 - x_f = x_s \ln\frac{x_0}{x_f} . \tag{6.38}$$

The total pulse energy delivered by stimulated emission from the beginning to the end is equal to $\hbar\omega(\Delta N^{(0)} - \Delta N^{(f)})/2$. The decay constant due to the output coupling of the resonator is denoted by κ_T as before, and the output pulse energy available from the Q-switched laser is given by

$$W = \frac{\kappa_T}{\kappa} \cdot \frac{\hbar\omega}{2} (\Delta N^{(0)} - \Delta N^{(f)}) . \tag{6.39}$$

On differentiating (6.36) with respect to x, we obtain

$$\frac{dy}{dx} = -\frac{1}{2} + \frac{x_s}{2x} ,$$

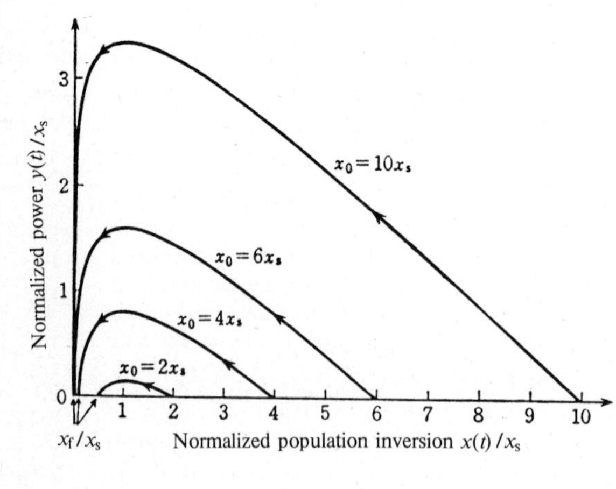

Fig. 6.9. Relation between the normalized power output $y(t)$ and the relative population inversion $x(t)/x_s$ of a Q-switched laser. x_0 is the initial value of the population inversion. x_s is the population inversion under steady-state oscillation

from which we find that the output power y becomes a maximum when $x = x_s$ (the threshold value), as mentioned above. The peak power of the output pulse normalized to the steady-state power then becomes

$$y_{\max} = \frac{x_0 - x_s}{2} - \frac{x_s}{2} \ln \frac{x_0}{x_s} . \tag{6.40}$$

Therefore, the peak power of the Q-switched laser is

$$\begin{aligned} P_{\max} &= 2\kappa_T \frac{y_{\max}}{\tau B(\omega)} \\ &= \hbar\omega\kappa_T \left(\Delta N^{(0)} - \Delta N_{\text{th}} - \Delta N_{\text{th}} \ln \frac{\Delta N^{(0)}}{\Delta N_{\text{th}}} \right) . \end{aligned} \tag{6.41}$$

If the pulse energy can be expressed by the product of the peak power and the effective pulse width Δt, we have

$$\Delta t = \frac{W}{P_{\max}} = \frac{1}{2\kappa} \left(1 - \frac{1}{\mathcal{N} - 1} \ln \mathcal{N} \right)^{-1} , \quad \text{where} \tag{6.42}$$

$$\mathcal{N} = \frac{x_0}{x_s} = \frac{\Delta N^{(0)}}{\Delta N_{\text{th}}}$$

is the relative excitation just when Q is switched.

In order to understand further characteristics of the Q-switched laser some typical numerical values of the parameters are shown below. The lifetime of atoms in the upper level of a solid-state laser, such as the ruby, YAG, or Nd-glass lasers, is about $\tau = 10^{-4}$–10^{-3} s. Before Q is switched at $t = 0$, the loss in the resonator is so large that light in the resonator may decay in the time of a few round-trips between the mirrors. Hence the optical decay constant is $\kappa' \cong 10^9\,\text{s}^{-1}$ initially. Then the threshold value $2\kappa'\tau$ is so high while $t < 0$, that the population inversion x_0 can be increased to a value comparable to $\kappa'\tau = 10^5$–10^6 under continuous pumping without building up oscillation. Raising the Q of the resonator suddenly at $t = 0$ so as to reduce κ to below $10^7\,\text{s}^{-1}$ for $t > 0$ and to lower the threshold $x_s = 2\kappa\tau$ to less than one hundredth of x_0, we have $x_s \ll x_0$, and (6.40) can be approximated to be

$$y_{\max} \approx \frac{\chi_0}{2}$$

within an error of a few percent.

From (6.41) the peak power of the laser output can then be expressed as

$$P_{\max} \approx \hbar\omega\kappa_T \Delta N^{(0)} . \tag{6.43}$$

Since x_f is smaller than x_s, whereas x_0 is about one hundred times larger than x_s, the approximation $x_f \ll x_0$ may be used to obtain the output energy from (6.39), with the result

$$W \approx \frac{\hbar\omega}{2} \cdot \frac{\kappa_T}{\kappa} \Delta N^{(0)} . \tag{6.44}$$

Then, the effective pulse width is given by

$$\Delta t \approx \frac{1}{2\kappa} . \tag{6.45}$$

Since it is not possible in steady-state operation to make the population inversion much greater than the threshold value, the steady-state output power from (6.15), with $\Delta N^{(0)} \approx 2\Delta N_{th}$, for example, becomes

$$P_{out} \approx \frac{\hbar\omega\kappa_T}{2\kappa\tau} \Delta N_{th} . \tag{6.46}$$

In the Q-switched laser, on the other hand, $\Delta N^{(0)}$ can be made as large as one hundred times ΔN_{th}, and $\kappa\tau$ becomes of the order 10^3–10^4 when $\tau = 10^{-4}$–10^{-3} s and $\kappa = 10^7\,s^{-1}$. Therefore, it may be seen that the peak power (6.43) of the Q-switched laser is 10^4–10^5 times that of the steady-state power (6.46). For example, if we apply Q-switching to a laser whose steady-state output is 1 kW, we can obtain a pulse with a peak power of 10–100 MW.

The peak output power is calculated from (6.43), when the population inversion is $\Delta N^{(0)} = 10^{19}\,cm^{-3}$, with $\kappa_T = \kappa/2 = 5 \times 10^6\,s^{-1}$ and $\hbar\omega = 2 \times 10^{-19}$ J, to become $P_{max} = 10^7\,W\,cm^{-3}$. Again, by putting $\kappa = 10^7\,s^{-1}$, the pulse width becomes $\Delta t = 50$ ns from (6.45). If κ is made less than $10^7\,s^{-1}$, the pulse width becomes broader in inverse proportion to κ. But the pulse energy is unchanged if the ratio of κ_T to κ is kept constant, and the peak power becomes smaller in proportion to κ. On the other hand, if κ is made larger, the pulse width becomes narrower and the peak power becomes greater. However, if κ is made several times larger than $10^7\,s^{-1}$ and the pulse width is made less than or about 10 ns, the threshold value $x_s = 2\kappa\tau$ becomes greater in proportion to κ, and the approximation $x_s \ll x_0$ will no longer be valid; therefore, both the peak power and the pulse energy will be less than the values given by (6.43, 44). Thus, in practice, the lower limit for the width of the giant pulse is about 10 ns.

When the light intensity changes very rapidly, such parameters as the transition probability and the relaxation rate are different from those in the quasi-steady state, so that the fundamental assumptions of the rate equations may no longer be valid. However, in the case of most solid-state lasers the rate-equation approximation is often found to be satisfactory on a time scale of about 1 ns, while for semiconductor lasers it is considered to be satisfactory for a time scale as short as 10 ps. More rigorous treatments must resort to the semi-classical theory given in Chap. 9.

Problems

6.1 The effect of spontaneous emission was neglected in (6.3). Find the term to be added on the right-hand side of (6.3) that represents the effect of spontaneous emission.

Hint: The Einstein A coefficient is the probability of emitting a photon regardless of the modes in space.
Answer: $(\hbar\omega)^2 N_2 B(\omega)$.

6.2 Find a formula for the threshold population inversion as a function of frequency, assuming that the gain profile is Lorentzian.

Answer: $$\Delta N_{\mathrm{th}}(\omega) = \frac{2\pi\kappa}{\hbar B}\left[\frac{(\omega-\omega_0)^2 + (\Delta\omega)^2}{\omega\Delta\omega}\right].$$

6.3 Draw a curve showing the frequency dependence of the steady-state power of a laser, having a Lorentzian gain profile.

Answer: Using the result of Problem 6.2, we find

$$P_{\mathrm{ss}} = \frac{\hbar\omega\Delta N_{\mathrm{th}}(\omega_0)}{2\tau}\left[\frac{\Delta N^{(0)}}{\Delta N_{\mathrm{th}}(\omega_0)} - 1 - \left(\frac{\omega-\omega_0}{\Delta\omega}\right)^2\right].$$

6.4 Verify that any tangent line on the curve for which P_{opt} against $\mathcal{N}$ is given by (6.20) is identical to the line representing (6.18).

6.5 Integrate (6.21) without using the approximation that $(1 + bW)^{-1} \cong 1 - bW$.

Answer: $2t = [a \ln(a - \kappa - b\kappa W_0)/(a - \kappa - b\kappa W) + \ln(W/W_0)]/(a - \kappa)$.

6.6 Draw a schematic diagram showing the relaxation oscillation given by (6.28, 29) in the x-y plane. If a microcomputer is available, plot the curve by numerical calculation for $x_0 = 2.5\,x_s$.

Hint: $dx/d\theta = 0$ at $y = (x_0 - x)/2x$, and $dy/d\theta = 0$ at $x = x_s$. The curve is a kind of converging spiral.

6.7 Find the full width at half maximum of the giant pulse in Fig. 6.8 and compare the value with Δt calculated from (6.42) under the same conditions.

Answer: FWHM $= 1.2\tau \times 10^{-4}$, $\Delta t = 1.34/2\kappa = 1.34\tau \times 10^{-4}$.

6.8 Find a formula for the optimum output coupling that gives the maximum pulse energy from the Q-switched laser given by (6.39).

Hint: $\Delta N_{\text{th}} = \Delta N_{\text{th, min}}(\kappa_0 + \kappa_T)/\kappa_0$. Use (6.17).
Answer:

$$\kappa_{T,\,\text{opt}} = \kappa_0 \left(\frac{\mathcal{N} - 1}{\ln \mathcal{N}} - 1 \right),$$

where $\mathcal{N}$ is defined by (6.37).

6.9 Find a formula for the ratio of the peak power of the giant pulse to the steady-state power as a function of $\mathcal{N} = x_0/x_{\text{s}}$. What is the limit of the ratio for $\mathcal{N} \to \infty$?

Answer: $y_{\max}/y_{\text{s}} = x_{\text{s}}[1 - (\mathcal{N} - 1)^{-1} \ln \mathcal{N}]$, $\lim\limits_{\mathcal{N} \to \infty} (y_{\max}/y_{\text{s}}) = x_{\text{s}} = 2\kappa\tau$.

7. Coherent Interaction

We have so far been concerned mainly with the transition probabilities of atoms and the optical energy emitted or absorbed in transitions. Here we must consider the coherent interaction between light and atoms in order to attain a basic understanding of the most important characteristic of the laser, namely, coherence. Then we discuss how the interaction of atoms with coherent light is different from an interaction with incoherent light.

When an atom is subjected to a coherent light field, it acquires a dipole moment of a certain fixed phase with respect to the optical field. Now, when many atoms interact with a common optical field, there exists a correlation between the phases of the induced atomic dipole moments. The atoms in such a state are said to have atomic coherence. Like the coherence of light, the degree of atomic coherence ranges from perfect coherence to complete incoherence. In this chapter we shall begin with the treatment of the case where there is practically no perturbation that destroys the atomic coherence.

7.1 Interaction Between a Two-Level Atom and a Coherent Field

Suppose that an atom has only two eigenstates, i.e., that there are only two energy levels[1]. Such a hypothetical atom is called a two-level atom. The wave functions of the eigenstates are the eigenfunctions $\psi_1(\boldsymbol{r},t)$ and $\psi_2(\boldsymbol{r},t)$ with eigenenergies W_1 and $W_2\,(>W_1)$, respectively. If the Hamiltonian of the atom is denoted by $\mathcal{H}$, the Schrödinger equation is written as

$$\mathrm{i}\hbar\frac{\partial\psi}{\partial t}=\mathcal{H}\psi\,. \tag{7.1}$$

In the absence of perturbation, the unperturbed Hamiltonian is denoted by $\mathcal{H}=\mathcal{H}_0$, and the solutions of (7.1) can be expressed as

$$\psi_n(\boldsymbol{r},t)=\phi_n(\boldsymbol{r})\exp\left[-\mathrm{i}\,(W_n/\hbar)\,t\right]\,, \tag{7.2}$$

$$\mathcal{H}_0\psi_n(\boldsymbol{r},t)=W_n\psi_n(\boldsymbol{r},t)\,, \tag{7.3}$$

1 The eigenstates are supposed to be non-degenerate so that there is only one upper state and one lower state.

where $n = 1$ and 2. The spatial part of the eigenfunction is denoted by $\phi_n(\boldsymbol{r})$. An unperturbed two-level atom is in either of the two eigenstates with an eigenenergy W_1 or W_2, and the electron distribution around the atomic nucleus is given by $|\psi_n(\boldsymbol{r}, t)|^2 = |\phi_n(\boldsymbol{r})|^2$.

Here, we shall assume that the wavelength of the optical field interacting with an atom is longer than the size of the atom. Therefore, the optical field on any one atom is considered to be spatially uniform and to change temporally with the frequency ω, as expressed by

$$E(t) = |\mathscr{E}| \cos(\omega t + \theta) = \tfrac{1}{2}(\mathscr{E} e^{i\omega t} + \mathscr{E}^* e^{-i\omega t}) , \tag{7.4}$$

where $\mathscr{E}$ is the complex amplitude given by

$$\mathscr{E} = |\mathscr{E}| e^{i\theta} .$$

Since $|\mathscr{E}|$ and θ are constant for a coherent field, the origin of time may be chosen so as to make $\theta = 0$.

In this chapter also, we shall consider only the electric dipole transition. Taking the z direction in the direction of polarization of the optical field and denoting the z-component of the electric dipole moment operator of the atom by μ_z, the perturbation Hamiltonian may be expressed as

$$\mathscr{H}'(t) = -\mu_z E(t) = -\tfrac{1}{2}\mu_z(\mathscr{E} e^{i\omega t} + \text{c.c.}) . \tag{7.5}$$

Since, in general, the wave function of a perturbed atom can be expressed by an expansion in terms of eigenfunctions, the wave function for an arbitrary state of the two-level atom can be written in the form

$$\psi(\boldsymbol{r}, t) = a_1(t)\,\psi_1(\boldsymbol{r}, t) + a_2(t)\,\psi_2(\boldsymbol{r}, t) , \tag{7.6}$$

where $a_1(t)$ and $a_2(t)$ represent the probability amplitudes of the eigenstates 1 and 2, respectively, and change relatively slowly with time. Substituting $\mathscr{H} = \mathscr{H}_0 + \mathscr{H}'(t)$ and (7.6) into the Schrödinger equation (7.1), we obtain

$$i\hbar\left(\frac{da_1}{dt}\psi_1 + a_1\frac{\partial\psi_1}{\partial t} + \frac{da_2}{dt}\psi_2 + a_2\frac{\partial\psi_2}{\partial t}\right) = (\mathscr{H}_0 + \mathscr{H}')(a_1\psi_1 + a_2\psi_2) . \tag{7.7}$$

Denoting the matrix elements of $\mathscr{H}'$ by

$$\mathscr{H}'_{nm} = \int \phi_n^* \mathscr{H}' \phi_m \, d\boldsymbol{r} , \tag{7.8}$$

we have $\mathscr{H}'_{11} = 0$ and $\mathscr{H}'_{22} = 0$, since the diagonal elements of μ_z are zero as described in detail in Sect. 4.4. Then operating with ψ_1^* on (7.7) and integrating spatially, we obtain

$$i\hbar\frac{da_1}{dt} = a_2\mathscr{H}'_{12} e^{-i\omega_0 t} , \tag{7.9 a}$$

where we have used the orthogonality of the eigenfunctions:

$$\int \psi_1^* \psi_2 \, d\boldsymbol{r} = \int \psi_2^* \psi_1 \, d\boldsymbol{r} = 0 \, .$$

Again, ω_0 is the proper frequency of the two-level atom given by

$$\omega_0 = \frac{W_2 - W_1}{\hbar} \, .$$

Similarly, operating ψ_2^* on (7.7) and integrating it, we obtain

$$\mathrm{i}\hbar \frac{da_2}{dt} = a_1 \mathcal{H}'_{21} \mathrm{e}^{\mathrm{i}\omega_0 t} \, . \tag{7.9b}$$

Now, if we substitute the matrix elements of the Hamiltonian of the electric dipole interaction (7.5),

$$\mathcal{H}'_{12} = -\tfrac{1}{2}\mu_{12} (\mathcal{E} \mathrm{e}^{\mathrm{i}\omega t} + \mathrm{c.\,c.}) \, , \quad \mathcal{H}'_{21} = -\tfrac{1}{2}\mu_{21} (\mathcal{E} \mathrm{e}^{\mathrm{i}\omega t} + \mathrm{c.\,c.}) \, , \tag{7.10}$$

into (7.9 a) and (7.9 b), respectively, we obtain

$$\begin{aligned} \frac{da_1}{dt} &= \frac{\mathrm{i}}{2\hbar} \mu_{12} a_2 (\mathcal{E} \mathrm{e}^{\mathrm{i}\omega t} + \mathrm{c.\,c.}) \mathrm{e}^{-\mathrm{i}\omega_0 t} \, , \\ \frac{da_2}{dt} &= \frac{\mathrm{i}}{2\hbar} \mu_{21} a_1 (\mathcal{E} \mathrm{e}^{\mathrm{i}\omega t} + \mathrm{c.\,c.}) \mathrm{e}^{\mathrm{i}\omega_0 t} \, . \end{aligned} \tag{7.11}$$

Since the term $\exp[\pm \mathrm{i}(\omega + \omega_0) t]$ on the right-hand side changes its sign very rapidly, it may be taken to be zero when averaged over a time scale longer than $1/\omega$. The approximation obtained by retaining the remaining resonance terms $\exp[\pm \mathrm{i}(\omega - \omega_0) t]$ is called the rotating-wave approximation. The meaning of this terminology will be made clear in Sect. 7.5. Under the rotating-wave approximation (7.11) becomes

$$\frac{da_1}{dt} = \frac{\mathrm{i}x}{2} a_2 \mathrm{e}^{\mathrm{i}(\omega - \omega_0)t} \, , \quad \frac{da_2}{dt} = \frac{\mathrm{i}x^*}{2} a_1 \mathrm{e}^{-\mathrm{i}(\omega - \omega_0)t} \, , \tag{7.12}$$

where we have put

$$x = \frac{\mu_{12} \mathcal{E}}{\hbar} \, , \quad x^* = \frac{\mu_{21} \mathcal{E}^*}{\hbar} \, .$$

The simultaneous differential equations (7.12) can be easily solved when x and ω of the incident light field can be assumed constant. Thus, we differentiate the first equation of (7.12) with respect to t and substitute it into the second equation to obtain

$$\frac{d^2 a_1}{dt^2} - \mathrm{i}(\omega - \omega_0) \frac{da_1}{dt} + \frac{|x|^2}{4} a_1 = 0 \, . \tag{7.13}$$

The solution of this equation may be generally written as

$$a_1(t) = (A_1 \mathrm{e}^{\mathrm{i}\Omega t/2} + B_1 \mathrm{e}^{-\mathrm{i}\Omega t/2})\, \mathrm{e}^{\mathrm{i}(\omega-\omega_0)t/2}\,, \tag{7.14}$$

$$\Omega = \sqrt{(\omega-\omega_0)^2 + |x|^2}\,, \tag{7.15}$$

where A_1 and B_1 are integration constants determined from the initial conditions.

Now, suppose that initially ($t < 0$) the atom is in the upper level 2, and that it is subjected to the coherent interaction given by (7.5) at $t = 0$ and thereafter. Then, the initial conditions are $a_1 = 0$ and $a_2 = 1$, with a result

$$A_1 = -B_1 = \frac{x}{2\Omega}\,,$$

so that, from (7.14, 12), we obtain

$$a_1(t) = \frac{\mathrm{i}x}{\Omega}\mathrm{e}^{\mathrm{i}(\omega-\omega_0)t/2}\sin\frac{\Omega t}{2}\,, \tag{7.16}$$

$$a_2(t) = \mathrm{e}^{-\mathrm{i}(\omega-\omega_0)t/2}\left(\cos\frac{\Omega t}{2} + \mathrm{i}\frac{\omega-\omega_0}{\Omega}\sin\frac{\Omega t}{2}\right). \tag{7.17}$$

Since the atom was initially in the upper level, $|a_1(t)|^2 = 1 - |a_2(t)|^2$ represents the probability of a transition to the lower level induced by the perturbation (7.5) during the time interval from $t = 0$ to t.

Since we have

$$|a_1(t)|^2 = \frac{|x|^2}{\Omega^2}\sin^2\frac{\Omega t}{2}\,, \tag{7.18}$$

the transition probability does not increase monotonically with time but increases and decreases periodically, as shown in Fig. 7.1.

It can be seen from Fig. 7.1 or (7.18) that during the time $t = 0$ to π/Ω the atom emits light, while during the time $t = \pi/\Omega$ to $2\pi/\Omega$ it absorbs light; after which, emission and absorption occur periodically. This is the case when

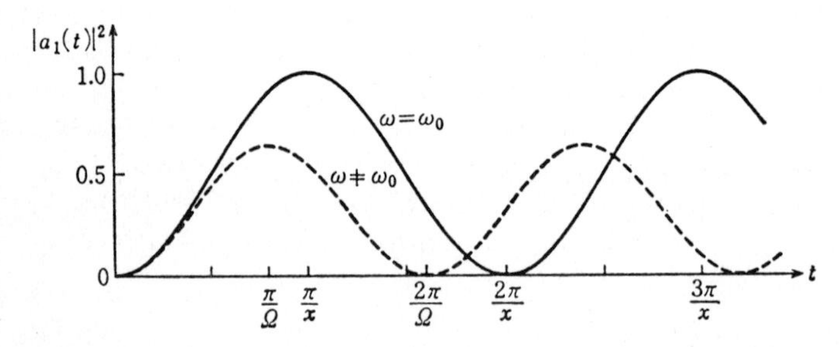

Fig. 7.1. Time evolution of the transition probability for a two-level atom undergoing coherent interaction

there is no perturbation other than the coherent interaction. In actual cases, however, there are relaxation effects such as collisions with other atoms and spontaneous emission. Therefore the sinusoidal curve in Fig. 7.1 is modified to a form of damped oscillation. The Ω of (7.15) is called the nutation frequency, and the value of Ω when $\omega = \omega_0$, namely, $|x| = |\mu_{12}\mathscr{E}|/\hbar$, is called Rabi's characteristic frequency, or the Rabi frequency, for short.

7.2 Induced Dipole Moment and Induced Emission Coefficient

As previously stated, a non-degenerate two-level atom in the stationary state ψ_1 or ψ_2 does not have any dipole moment, but, when subject to a coherent interaction, an oscillatory dipole moment $p(t)$ is induced. Writing e for the electron charge and taking the z axis in the direction of the optical field, $p(t)$ is given by the quantum-mechanical expectation value of ez as

$$p(t) = \int \psi^*(\boldsymbol{r}, t)\, ez\psi(\boldsymbol{r}, t)\, d\boldsymbol{r} . \tag{7.19}$$

Substituting (7.6) and (4.34) into this equation, we find

$$p(t) = a_2^* a_1 \mu_{21} \mathrm{e}^{\mathrm{i}\omega_0 t} + a_1^* a_2 \mu_{12} \mathrm{e}^{-\mathrm{i}\omega_0 t} . \tag{7.20}$$

Since a_1 and a_2 are given, respectively, by (7.16) and (7.17) for an atom perturbed by the coherent interaction (7.5), substitution of these into (7.20) gives

$$p(t) = \frac{\mathrm{i}x}{2\Omega}\mu_{21}\left(\sin\Omega t - \mathrm{i}\frac{\omega-\omega_0}{\Omega}(1-\cos\Omega t)\right)\mathrm{e}^{\mathrm{i}\omega t} + \mathrm{c.\,c.} \tag{7.21}$$

Although it appears in (7.20) that $p(t)$ oscillates with the proper frequency ω_0, it is seen in (7.21), where a_1 and a_2 have been substituted into (7.20), that it actually oscillates with the frequency ω of the incident light.

When $\omega = \omega_0$, the phase of the induced dipole moment is 90° ahead of the electric field during the initial stage $t = 0$ to π/Ω, as seen in (7.21). The polarization current $\partial p/\partial t$ is then 180° ahead in phase, namely, opposite in phase to the electric field. An electric current flowing in a direction opposite to the electric field is equivalent to a negative resistance; it absorbs negative energy from the electric field, namely, it supplies energy to the electric field. This is identical to the energy of induced emission. the power given to the electric field at $\omega = \omega_0$ by the dipole moment (7.21) is obtained from (7.21, 4):

$$P = \overline{\frac{\partial p}{\partial t}E} = \frac{\hbar\omega|x|^2}{2\Omega}\sin\Omega t , \tag{7.22}$$

which is equal to the time derivative of the energy $\hbar\omega|a_1(t)|^2$ of induced emission. This relation holds even when $\omega \neq \omega_0$: the energy lost by a two-level atom is, in general, equal to the work done by the induced dipole moment on the optical field.

Based on the considerations above, let us now think about the induced emission and absorption in the atom caused by incoherent light. When the atom is perturbed incoherently, the wave function may also be expressed as a linear superposition of ψ_1 and ψ_2 in the form of (7.6). However, the phases of the probability amplitudes a_1 and a_2 are indeterminate in this case. Even though $p(t)$ for each atom has a certain instantaneous value, the ensemble average for a large number of atoms or the time average over a long period is zero. But, since the average of $|p(t)|^2$ is not zero, the probability of induced transition is not zero, even in the case of incoherent light. Thus, we may still use the above calculation for the coherent interaction to obtain the Einstein B coefficient.

If the incident light is weak, $\mathscr{E}$ and hence x are so small that the transition probability given by (7.18) can be large only when the frequency ω is very close to ω_0. Needless to say, the absorption probability of an atom initially in the lower level is also equal to (7.18). Incoherent light has a continuous spectral distribution with the Fourier component $e(\omega)$ at the frequency ω. The statistical average of $e(\omega)$ vanishes but that of $|e(\omega)|^2$ does not vanish, and has a value which appears to be independent of ω in the neighborhood of ω_0. Thus, if we put

$$\left\langle \left| \frac{\mu_{12}\, e(\omega)}{\hbar} \right|^2 \right\rangle_{av} = x_{\text{eff}}^2 , \tag{7.23}$$

the probability of induced transition for this Fourier component becomes

$$|a(\omega, t)|^2 = \left(\frac{\sin(\omega - \omega_0)\, t/2}{\omega - \omega_0} \right)^2 x_{\text{eff}}^2$$

by putting $x \to 0$ in (7.18) and replacing $|x|^2$ by x_{eff}^2. By integrating the above expression from $t = 0$ to t, the probability that a two-level atom undergoes an induced transition during the time interval t is given by

$$\int_0^\infty |a(\omega, t)|^2 \, d\omega = \frac{\pi}{2} x_{\text{eff}}^2 t . \tag{7.24}$$

Since the optical electric field is $e(\omega)$ at a frequency band between ω and $\omega + d\omega$, its energy per unit volume is

$$\tfrac{1}{2}\, \varepsilon_0 |e(\omega)|^2 .$$

The statistical average of the energy density then becomes

$$\varrho(\omega) = \frac{\varepsilon_0 \hbar^2}{2|\mu_{12}|^2} x_{\mathrm{eff}}^2 . \tag{7.25}$$

Since the probability (7.24) of induced transition by light whose frequency lies between ω and $\omega + d\omega$ can be written as $B\varrho(\omega)t$, where B is the induced emission coefficient, it follows from (7.24, 25) that

$$\frac{\pi}{2} x_{\mathrm{eff}}^2 t = B \frac{\varepsilon_0 \hbar^2}{2|\mu_{12}|^2} x_{\mathrm{eff}}^2 t .$$

From this we obtain

$$B = \frac{\pi}{\varepsilon_0 \hbar^2} |\mu_{12}|^2 \tag{7.26}$$

which is equivalent to (4.36) obtained in Sect. 4.4.

In the above calculation, the z component (the component in the direction of the optical field) of the dipole moment has been denoted by μ. But, when the direction of each atom is random and isotropically distributed, by denoting the matrix element of the atomic dipole moment by μ_{a}, we have

$$B = \frac{\pi}{3\varepsilon_0 \hbar^2} |\mu_{\mathrm{a}}|^2 \tag{7.27}$$

since $\langle |\mu_{12}|^2 \rangle = |\mu_{\mathrm{a}}|^2/3$. This corresponds to the induced emission coefficient of a gaseous medium for linearly polarized light. Likewise, when the dipole moments are aligned, while the direction of polarization of the incident light is indeterminate, the induced emission coefficient is the same as (7.27). It is the same, also, when the direction of polarization of the incident light and that of the atoms are both indeterminate.

7.3 Density Matrix

In order to understand the interaction between an ensemble of atoms (or molecules) and light, it suffices to know the behavior of each atom and to find the ensemble average. However, such a calculation is far too complicated and often meets with unsurmountable mathematical difficulties. For example, it is not simple to deal even with the effects of such statistical phenomena as the molecular collisions in a gas.

Phenomena such as the response of a medium to incident light and the emission from an excited medium can be described in terms of the induced dipole moment, the transition probability, and the populations of atoms.

These can be expressed using $|a_1|^2$, $a_1 a_2^*$, $a_2 a_1^*$, and $|a_2|^2$ in the case of a two-level atom, and the probability amplitude alone does not appear in the expressions. Thus, by using a density matrix as defined below, various calculations of physical processes can often be simplified, and their physical meanings are made clearer.

It is, in general, possible to expand the wave function $\psi(\boldsymbol{r}, t)$ of a certain atom in terms of the eigenfunctions $\psi_n(\boldsymbol{r}, t)$ of the eigenstates in the form

$$\psi(\boldsymbol{r}, t) = \sum_n a_n(t)\,\psi_n(\boldsymbol{r}, t)\,. \tag{7.28}$$

The temporal function may be removed by rewriting (7.28) with the spatial wave function $\phi_n(\boldsymbol{r})$:

$$\psi(\boldsymbol{r}, t) = \sum_n c_n(t)\,\phi_n(\boldsymbol{r})\,, \quad \text{where} \tag{7.29}$$

$$c_n(t) = a_n(t)\,\mathrm{e}^{-\mathrm{i}(W_n/\hbar)t} \tag{7.30}$$

as can be immediately seen from (7.2). We use this to define the matrix elements of the density matrix ϱ as

$$\varrho_{nm} = c_n c_m^* \quad \text{or} \tag{7.31 a}$$

$$\varrho_{nm} = a_n a_m^* \exp\left(\mathrm{i}\frac{W_m - W_n}{\hbar}t\right)\,. \tag{7.31 b}$$

It is seen from (7.31) that

$$\varrho_{nm} = \varrho_{mn}^*\,, \tag{7.32}$$

so that the density matrix is Hermitian. Moreover, the normalization of the wave function is

$$\sum_n |a_n|^2 = \sum_n |c_n|^2 = 1\,,$$

which, for the density matrix, becomes

$$\mathrm{Tr}\{\varrho\} = \sum_n \varrho_{nn} = 1\,, \tag{7.33}$$

where Tr denotes the trace operator (the sum over the diagonal elements).

Using the density matrix, the expectation value of an operator A of a physical quantity is given by

$$\langle A\rangle = \int \psi^*(\boldsymbol{r}, t)\,A\,\psi(\boldsymbol{r}, t)\,d\boldsymbol{r} = \sum_n \sum_m c_n c_m^* A_{mn}\,,$$

so that

$$\langle A\rangle = \mathrm{Tr}\{\varrho A\} \tag{7.34}$$

because

$$\sum_m \varrho_{nm} A_{mn} = (\varrho A)_{nn}\,.$$

The density matrix of the two-level atom discussed in Sect. 7.1 is expressed as

$$\varrho = \begin{bmatrix} |a_1|^2 & a_1 a_2^* \, \mathrm{e}^{\mathrm{i}\omega_0 t} \\ a_2 a_1^* \, \mathrm{e}^{-\mathrm{i}\omega_0 t} & |a_2|^2 \end{bmatrix},$$

so that the diagonal elements give, indeed, the population (probability of the number) of atoms and the off-diagonal elements give the complex dipole moment. We may calculate the expectation value of the dipole moment operator

$$\begin{pmatrix} \mu_{11} & \mu_{12} \\ \mu_{21} & \mu_{22} \end{pmatrix}$$

by using (7.34) to obtain (7.20).

In this description of the density matrix we have assumed that the atom is in a quantum-mechanically pure state and that the wave function $\psi(\boldsymbol{r}, t)$ is definite, since the probability amplitudes $a_n(t)$ are known. However, the density matrix is also applicable to the case of what is called a mixed state, where the wave function of each atom in the atomic ensemble is not known but only their statistical average. Since the probability amplitude of each atom is indeterminate in the mixed state, the density matrix of each atom is also indeterminate; we may, however, consider the ensemble average $\langle \varrho \rangle_{\mathrm{av}}$ to be a definite function which varies continuously. Using the density matrix of an ensemble average, it will be made clear from the following that the expectation value of any physical quantity of that atomic ensemble is given by

$$\langle A \rangle_{\mathrm{av}} = \mathrm{Tr}\{\langle \varrho \rangle_{\mathrm{av}} A\}, \tag{7.35}$$

a similar form to (7.34).

The probability amplitude $a_n(t)$ of each atom in a mixed state is indefinite, but the statistical probability $p^{(j)}$ that its wave function takes a pure state

$$\psi^{(j)} = \sum_n a_n^{(j)} \psi_n$$

is known. The matrix element of the density matrix $\langle \varrho \rangle_{\mathrm{av}}$ of the ensemble average is then found to be

$$\begin{aligned} (\langle \varrho \rangle_{\mathrm{av}})_{nm} &= \langle c_n c_m^* \rangle_{\mathrm{av}} = \sum_j p^{(j)} c_n^{(j)} c_m^{(j)*} \\ &= \sum_j p^{(j)} a_n^{(j)} a_m^{(j)*} \, \mathrm{e}^{-\mathrm{i}\omega_{nm} t} \end{aligned} \tag{7.36}$$

where $\hbar\omega_{nm} = W_n - W_m$ is the difference of the eigenenergies. The statistical average of the expectation value of an operator A is

$$\langle A \rangle_{\mathrm{av}} = \sum_j p^{(j)} \int \psi^{(j)*} A \psi^{(j)} \, d\boldsymbol{r}$$

which is rewritten using (7.36) as

$$\langle A\rangle_{\mathrm{av}} = \sum_j p^{(j)} \sum_n \sum_m c_n^{(j)} c_m^{(j)*} A_{mn}$$
$$= \sum_j p^{(j)} \sum_n \sum_m \varrho_{nm} A_{mn} = \sum_n \sum_m \left(\sum_j p^{(j)} \varrho_{nm}\right) A_{mn}$$
$$= \sum_n \sum_m (\langle\varrho\rangle_{\mathrm{av}})_{nm} A_{mn} = \mathrm{Tr}\,\{\langle\varrho\rangle_{\mathrm{av}} A\} \ .$$

Thus we have obtained (7.35).

7.4 Equations of Motion of the Density Matrix

Equations of motion describing the temporal variation of the density matrix can be obtained from the Schrödinger equation (7.1). The time rate of change of c_n is obtained by substituting (7.29) into (7.1), as in the calculation for the two-level atom in Sect. 7.1. When the Hamiltonian is $\mathcal{H} = \mathcal{H}_0 + \mathcal{H}'$, where $\mathcal{H}'$ is the perturbation, in general, we have

$$i\hbar \frac{dc_n}{dt} = \sum_k c_k \mathcal{H}_{nk} \ . \tag{7.37}$$

To obtain the temporal change of the density matrix, the matrix elements (7.31 a) are differentiated to give

$$\frac{d\varrho_{nm}}{dt} = \frac{dc_n}{dt} c_m^* + c_n \frac{dc_m^*}{dt} \ .$$

Substituting (7.37) and its complex conjugate into the above equation, we obtain, since $\mathcal{H}_{nk}^* = \mathcal{H}_{kn}$,

$$i\hbar \frac{d\varrho_{nm}}{dt} = \sum_k c_k \mathcal{H}_{nk} c_m^* - c_n \sum_k c_k^* \mathcal{H}_{km}$$
$$= \sum_k (\mathcal{H}_{nk} \varrho_{km} - \varrho_{nk} \mathcal{H}_{km}) \ . \tag{7.38}$$

By using the commutator notation

$$[\mathcal{H}, \varrho] = \mathcal{H}\varrho - \varrho\mathcal{H} \ ,$$

(7.38) can be written as

$$i\hbar \frac{d\varrho}{dt} = [\mathcal{H}, \varrho] \ , \tag{7.39}$$

which is usually called the equation of motion of the density matrix. Denoting the eigenvalue of the unperturbed Hamiltonian $\mathcal{H}_0$ by W_n as in (7.3), we rewrite (7.38) as

$$i\hbar \frac{d\varrho_{nm}}{dt} = W_n\varrho_{nm} - W_m\varrho_{nm} + \sum_k (\mathcal{H}'_{nk}\varrho_{km} - \varrho_{nk}\mathcal{H}'_{km})$$

or

$$\frac{d\varrho_{nm}}{dt} = -i\omega_{nm}\varrho_{nm} - \frac{i}{\hbar}\sum_k (\mathcal{H}'_{nk}\varrho_{km} - \varrho_{nk}\mathcal{H}'_{km}) . \tag{7.40}$$

This equation is often employed in problem-solving.

The advantage of using the density matrix is that the same equation of motion (7.39) can be used for mixed states and for pure states. In many-body problems, it is often the case that the statistical average of the density matrix is known, although the wave functions of the individual particles are unknown. By using the density matrix in such cases, any physical quantity can be calculated in a manner similar to a calculation for a pure state. Since no difficulty arises in denoting the density matrix of the mixed state $\langle\varrho\rangle_{\mathrm{av}}$ by ϱ, in common with that of the pure state, we shall simply use ϱ in either case.

In Sect. 7.1 we investigated the behavior of a two-level atom perturbed only by a coherent interaction. In an ensemble of two-level atoms, not only are there interactions between the atoms, but there are also spontaneous emission and incoherent interactions with other atoms in the medium, such as impurities, walls of the container, atoms of the host crystal in the case of a solid, and solvent molecules in the case of a liquid. These interactions have no phase relation with the incident wave. It is, in practice, impossible to trace the variation of the wave function of every atom under such incoherent perturbations. However, by dealing with them using the density matrix, it is simple to introduce the effects of relaxation and decay into the theory in a so-called phenomenological approximation. Since most phenomena related to lasers may be treated with the two-level-atom approximation, we shall hereafter discuss only the density matrix of a two-level atom for the sake of simplicity and clarity. Nevertheless, it is easy to extend the following to the case of a many-level atom.

When, in general, a two-level atom with an energy-level separation of $\hbar\omega_0 = W_2 - W_1$, is subject to a perturbation $\mathcal{H}'$, the equation of motion of the density matrix with two rows and two columns can be written in terms of the matrix elements from (7.40) as

$$\frac{d\varrho_{11}}{dt} = \frac{i}{\hbar}(\varrho_{12}\mathcal{H}'_{21} - \mathrm{c.c.}) , \tag{7.41}$$

$$\frac{d\varrho_{12}}{dt} = i\omega_0\varrho_{12} - \frac{i}{\hbar}(\varrho_{22} - \varrho_{11})\mathcal{H}'_{12} . \tag{7.42}$$

The equations for the remaining elements can be obtained from these, since $\varrho_{21} = \varrho^*_{12}$ and $\varrho_{22} = 1 - \varrho_{11}$. Here, ϱ_{11} and ϱ_{22} represent the populations of the atom, while the dipole moment (7.20) is given by

$$\langle p(t)\rangle_{\mathrm{av}} = \varrho_{12}\mu_{21} + \mathrm{c.c.} \tag{7.43}$$

Hence it may be said that ϱ_{12} is the relative value of the complex dipole moment, or the normalized dipole moment.

When there are n atoms in a volume $V(n \gg 1)$, the polarization of the medium of two-level atoms is

$$P = \frac{1}{V}\sum_{k=1}^{n} p_k = N\langle p\rangle \tag{7.44}$$

where p_k is the dipole moment of the kth atom, $N = n/V$ is the number of atoms per unit volume, and $\langle p\rangle$ is the ensemble average of the dipole moments. Again, denoting the density of population inversion $\varrho_{22} - \varrho_{11}$ by $\Delta\varrho$, the population inversion in the number of atoms is

$$\Delta N = (\varrho_{22} - \varrho_{11})N = N\Delta\varrho \ .$$

The change of $\Delta\varrho$ with time, from (7.41), becomes

$$\frac{d}{dt}\Delta\varrho = -\frac{2\mathrm{i}}{\hbar}(\varrho_{12}\mathcal{H}'_{21} - \mathrm{c.c.}) \ .$$

By separating the perturbations of the two-level atom into the coherent perturbation of the laser field and incoherent perturbations such as those due to excitation and relaxation, we shall hereafter restrict the notation $\mathcal{H}'$ only to the coherent perturbation, while assuming that the incoherent perturbations are expressed by the coefficient of excitation Γ_{ex} and by the relaxation time τ. Then the above equation will be replaced by

$$\frac{d}{dt}\Delta\varrho = \Gamma_{\mathrm{ex}} - \frac{\Delta\varrho}{\tau} - \frac{2\mathrm{i}}{\hbar}(\varrho_{12}\mathcal{H}'_{21} - \mathrm{c.c.}) \ . \tag{7.45}$$

Although this equation appears to be phenomenologically correct, it is an approximation which is valid only when the time scale is longer than the characteristic time of excitation or relaxation (time interval between collisions of molecules in the case of a gas) from a microscopic point of view. In the ordinary theory of lasers, however, we may use the above equation as one of the fundamental equations in practically all cases.

Now, there are two opposing processes in excitation: one is the pumping, so named because the atoms in the lower level are pumped up to the upper level so as to create a population inversion through irradiation with strong light, or through electron impact, etc., and the other is the thermal de-excitation which reduces the population inversion through thermal agitation. Then, the rate of excitation is expressed as

$$\Gamma_{\mathrm{ex}} = \Gamma_{\mathrm{p}} - \Gamma' \ ,$$

where Γ_{p} is the rate of pumping and Γ' is the rate of thermal de-excitation.

When there is neither pumping nor a laser field, i.e., $\Gamma_p = \mathscr{H}' = 0$, we obtain from (7.45) in the steady state

$$\Gamma' = -\frac{\Delta\varrho^{(T)}}{\tau} = \frac{1}{\tau}\left(\varrho_{11}^{(T)} - \varrho_{22}^{(T)}\right), \tag{7.46}$$

where $\varrho_{11}^{(T)}$ and $\varrho_{22}^{(T)}$ are populations in thermal equilibrium[2]. Since they must satisfy the Boltzmann distribution, we have

$$\varrho_{22}^{(T)} = \varrho_{11}^{(T)} \mathrm{e}^{-\hbar\omega_0/k_B T} .$$

Substituting this into (7.46), and since $\varrho_{22}^{(T)} = 1 - \varrho_{11}^{(T)}$,

$$-\Delta\varrho^{(T)} = \tau\Gamma' = \tanh\left(\frac{\hbar\omega_0}{2k_B T}\right) .$$

Since $\hbar\omega_0 \ll k_B T$ for a transition at a long wavelength, we may usually take $\tau\Gamma' \approx 0$. But more generally, if we include the term Γ', and accordingly $\Delta\varrho^{(T)}$ (< 0), (7.45) may be rewritten

$$\frac{d}{dt}\Delta\varrho = \Gamma_p - \frac{\Delta\varrho - \Delta\varrho^{(T)}}{\tau} - \frac{2\mathrm{i}}{\hbar}\left(\varrho_{12}\mathscr{H}'_{21} - \mathrm{c.c.}\right) . \tag{7.47}$$

Instead of using the rate of pumping Γ_p, it is more common to use the unperturbed population inversion $\Delta\varrho^{(0)}$, which is the population difference of the two-level atoms under pumping in the absence of perturbation by the laser field. From (7.45), when $\Gamma_p \neq 0$ and $\mathscr{H}' = 0$, we have

$$\Delta\varrho^{(0)} = \tau\Gamma_{ex} = \tau\Gamma_p - \tau\Gamma' .$$

Using (7.46), this becomes

$$\Delta\varrho^{(0)} = \tau\Gamma_p + \Delta\varrho^{(T)} .$$

Thus (7.45 or 47) for $\mathscr{H}' \neq 0$ may generally be modified to

$$\frac{d}{dt}\Delta\varrho = -\frac{\Delta\varrho - \Delta\varrho^{(0)}}{\tau} - \frac{2\mathrm{i}}{\hbar}\left(\varrho_{12}\mathscr{H}'_{21} - \mathrm{c.c.}\right) . \tag{7.48}$$

Next, we denote the relaxation time of the polarization, i.e., the ensemble average of the atomic dipole moments, by $1/\gamma$ and modify (7.42) to

$$\frac{d}{dt}\varrho_{12} = (\mathrm{i}\omega_0 - \gamma)\varrho_{12} - \frac{\mathrm{i}}{\hbar}\left(\varrho_{22} - \varrho_{11}\right)\mathscr{H}'_{12} . \tag{7.49}$$

it should be noticed that τ and $1/\gamma$ are not usually equal to one another.

---

2 The superscript (T) indicates thermal equilibrium at the temperature T.

Following the terminology used in magnetic resonance, τ is called the longitudinal relaxation time and $1/\gamma$ the transverse relaxation time. In magnetic resonance they are customarily denoted by T_1 and T_2, respectively, but here, to avoid confusion with the notation for energy levels, we use τ and $1/\gamma$. Then $1/\tau$ is the longitudinal relaxation constant and γ the transverse relaxation constant.

7.5 Optical Bloch Equations

Problems on coherent interaction when there is no relaxation (or when the relaxation is considered as microscopic elementary processes) can be solved by using (7.9 a and b), whereas those with phenomenological relaxation constants can be solved by using (7.49) and (7.45 or 48) with explicit forms for the perturbations and initial conditions. Effects of the coherent interaction in the optical region may also be represented geometrically by using a precessing vector model analogous to the precession and nutation of magnetic moments, well known in microwave and radiofrequency magnetic resonance. This model was introduced by *R. P. Feynman* et al. [7.1] and is useful in presenting an intuitive view of the problems. By the transformation given below, the coherent electric dipole interaction between the two-level atom and the optical field can be expressed in the form of Bloch equations describing the magnetic moment in a magnetic field. Hence, they have come to be called the optical Bloch equations.

7.5.1 Transformation to an Abstract Space

Besides the physical space (x, y, z) of reality, a vector $\boldsymbol{\varrho}$ in an abstract three-dimensional space (X, Y, Z) is considered to have three components, defined by

$$\begin{aligned} \varrho_X &\equiv \varrho_{12} + \varrho_{21} = 2\,\mathrm{Re}\{\varrho_{12}\}\,, \\ \varrho_Y &\equiv \frac{1}{\mathrm{i}}(\varrho_{12} - \varrho_{21}) = 2\,\mathrm{Im}\{\varrho_{12}\}\,, \\ \varrho_Z &\equiv \varrho_{22} - \varrho_{11} = \Delta\varrho\,. \end{aligned} \tag{7.50}$$

In terms of this vector $\boldsymbol{\varrho} = (\varrho_X, \varrho_Y, \varrho_Z)$, often called a pseudovector or an optical Bloch vector, (7.41, 42) become

$$\frac{d}{dt}\varrho_X = -\omega_0\varrho_Y + \frac{1}{\mathrm{i}\hbar}(\mathscr{H}'_{12} - \mathscr{H}'_{21})\,\varrho_Z\,, \tag{7.51 a}$$

$$\frac{d}{dt}\varrho_Y = \omega_0\varrho_X - \frac{1}{\hbar}(\mathcal{H}'_{12} + \mathcal{H}'_{21})\varrho_Z , \tag{7.51 b}$$

$$\frac{d}{dt}\varrho_Z = -\frac{1}{i\hbar}(\mathcal{H}'_{12} - \mathcal{H}'_{21})\varrho_X + \frac{1}{\hbar}(\mathcal{H}'_{12} + \mathcal{H}'_{21})\varrho_Y . \tag{7.51 c}$$

Now imagine a fictitious force $\boldsymbol{F} = (F_X, F_Y, F_Z)$ whose components in this abstract space are

$$F_X = \frac{1}{\hbar}(\mathcal{H}'_{12} + \mathcal{H}'_{21}) ,$$

$$F_Y = \frac{1}{i\hbar}(\mathcal{H}'_{12} - \mathcal{H}'_{21}) , \tag{7.52}$$

$$F_Z = \omega_0 .$$

Then (7.51 a–c) can be simplified to

$$\frac{d\boldsymbol{\varrho}}{dt} = \boldsymbol{F} \times \boldsymbol{\varrho} . \tag{7.53}$$

This is of exactly the same form as the vector equation of motion of a magnetic moment in a magnetic field or a top spinning in a gravitational field.

Let us first investigate the motion of $\boldsymbol{\varrho}$ in the absence of perturbation. Since $F_X = F_Y = 0$ when $\mathcal{H}' = 0$, from (7.52), $d\varrho_Z/dt = 0$ in (7.51 c or 53), so that the Z component of $\boldsymbol{\varrho}$ is constant. The component of $\boldsymbol{\varrho}$ in the XY plane rotates with angular velocity ω_0, as shown in Fig. 7.2. This corresponds to the precession of a spinning top or a magnetic moment. The ω_0 of the two-level atom corresponds to the gravitational force for the top or the static magnetic field for the magnetic moment.

In the presence of perturbation, the correspondence between magnetic resonance in real space (x, y, z) and the two-level atom in the abstract space (X, Y, Z) is such that H_z corresponds to ω_0 as mentioned above, whereas the x and y components of the perturbing magnetic field correspond to the real and imaginary parts of $\mathcal{H}'_{12}$, respectively. In other words, the perturbation $\mathcal{H}'_{12}$ due to the optical field is represented by a complex vector in the XY plane. The perturbation given by the coherent optical field $E(t)$ of (7.4) is

$$\mathcal{H}'_{12} = -\mu_{12}E(t) = -\tfrac{1}{2}\mu_{12}(\mathcal{E}e^{i\omega t} + \mathcal{E}^* e^{-i\omega t}) \tag{7.54}$$

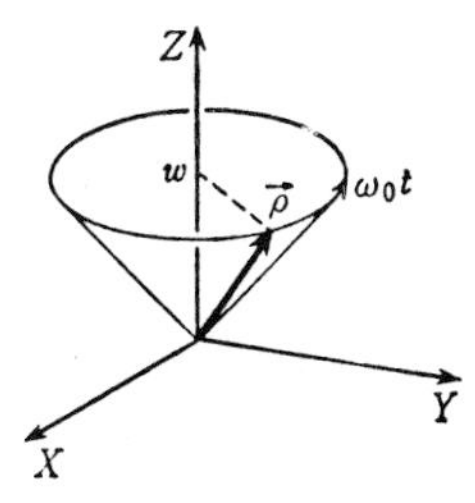

Fig. 7.2. Precessional motion of an optical Bloch vector

and its complex representation in the XY plane consists of two terms: one has $\exp(\mathrm{i}\omega t)$, showing a left-handed rotating wave, and the other has $\exp(-\mathrm{i}\omega t)$, showing a right-handed rotating wave. Thus, if $\omega \approx \omega_0$, the former is in resonance with a vector $\boldsymbol{\varrho}$ precessing with the angular velocity ω_0 about the Z axis, while the latter will not be in resonance because it is rotating in the opposite direction. The reason why the approximation (7.12) in Sect. 7.1 is called the rotating-wave approximation is because it is an approximation retaining only the perturbation rotating in the same direction as the rotation of $\boldsymbol{\varrho}$ in the XY plane.

7.5.2 Representation in a Rotating Frame of Reference

The concept of the rotating-wave approximation will be even clearer if we transform the vector into a frame of reference (X', Y', Z') which is rotating with an angular velocity ω about the Z axis (Fig. 7.3). According to Larmor's theorem, the time derivative $\partial/\partial t$ in the rotating frame and the time derivative d/dt in the stationary frame are related as

$$\frac{\partial \boldsymbol{\varrho}}{\partial t} = \frac{d\boldsymbol{\varrho}}{dt} - \boldsymbol{\omega} \times \boldsymbol{\varrho} \ , \tag{7.55}$$

where the components of the vector $\boldsymbol{\omega}$ are $(0, 0, \omega)$. Then (7.53) in the rotating frame becomes

$$\frac{\partial \boldsymbol{\varrho}}{\partial t} = (\boldsymbol{F} - \boldsymbol{\omega}) \times \boldsymbol{\varrho} \ . \tag{7.56}$$

The Z component of $\boldsymbol{F} - \boldsymbol{\omega}$ is $\omega_0 - \omega$, while the X and Y components in the stationary frame are $2\,\mathrm{Re}\{\mathscr{H}'_{12}/\hbar\}$ and $2\,\mathrm{Im}\{\mathscr{H}'_{12}/\hbar\}$, respectively, from (7.52). Thus, when $\omega = \omega_0$, the first term of (7.54) in the rotating frame is a complex vector stationary with respect to the frame, and the second term is a complex vector rotating with an angular velocity of -2ω.

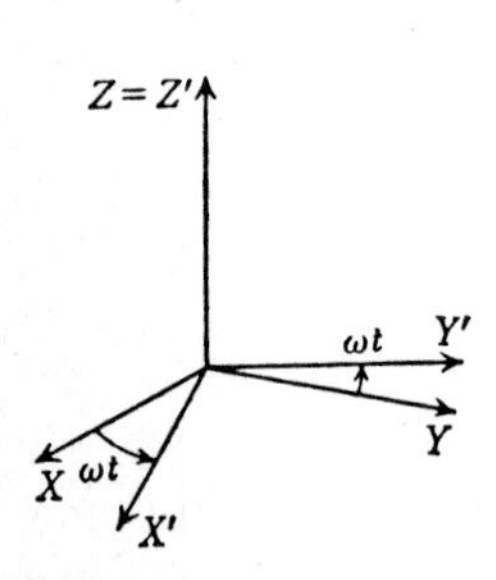

Fig. 7.3. Frame of reference (X', Y', Z') rotating with angular velocity ω about the Z axis of the stationary frame (X, Y, Z)

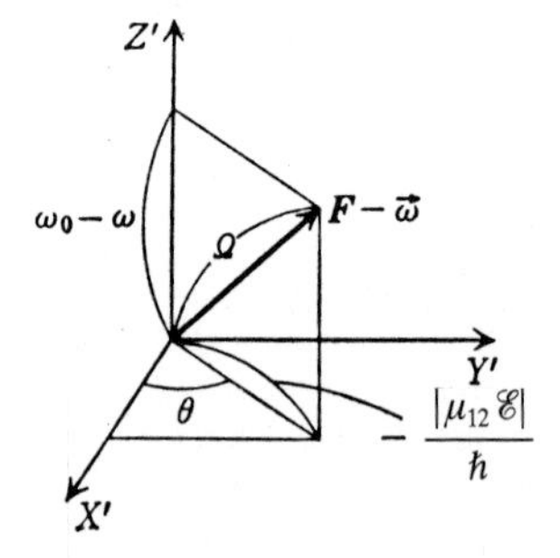

Fig. 7.4. Perturbation vector represented in the rotating frame

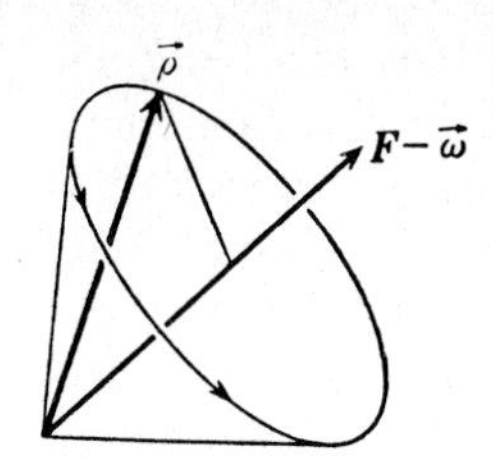

Fig. 7.5. Motion of the Bloch vector $\boldsymbol{\varrho}$ precessing around the perturbation vector $\boldsymbol{F} - \boldsymbol{\omega}$

If we neglect the second term in (7.54) in the rotating-wave approximation, the perturbation $\mathscr{H}'_{12} = -\frac{1}{2}\mu_{12}\mathscr{E}\exp(\mathrm{i}\omega t)$ becomes $\mathscr{H}'_{12} = -\frac{1}{2}\mu_{12}\mathscr{E}$ in the rotating frame, so that

$$F_{X'} = \frac{2}{\hbar}\mathrm{Re}\{\mathscr{H}'_{12}\} = -\frac{1}{\hbar}\mathrm{Re}\{\mu_{12}\mathscr{E}\}\,, \tag{7.57 a}$$

$$F_{Y'} = \frac{2}{\hbar}\mathrm{Im}\{\mathscr{H}'_{12}\} = -\frac{1}{\hbar}\mathrm{Im}\{\mu_{12}\mathscr{E}\}\,. \tag{7.57 b}$$

Therefore, the absolute value of the complex vector representing the projection of $\boldsymbol{F}$ or $\boldsymbol{F} - \boldsymbol{\omega}$ on the $X'Y'$ plane is $-|\mu_{12}\mathscr{E}|/\hbar$, shown in Fig. 7.4, in which the argument θ represents the phase angle θ of the incident wave (7.4).

When viewed in the rotating frame, (7.56) indicates that the motion of $\boldsymbol{\varrho}$ is a slow precession with angular velocity $|\boldsymbol{F} - \boldsymbol{\omega}|$ about the vector $\boldsymbol{F} - \boldsymbol{\omega}$, as shown in Fig. 7.5. In order to distinguish it from the rapid precession ω about the Z axis that is seen in the stationary frame, this motion is called nutation. The resulting motion of $\boldsymbol{\varrho}$ under a perturbation in the stationary frame is a superposition of the nutation on the precession. It is immediately clear from Fig. 7.4 that the angular velocity of nutation is

$$|\boldsymbol{F} - \boldsymbol{\omega}| = \sqrt{\left|\frac{\mu_{12}\mathscr{E}}{\hbar}\right|^2 + (\omega_0 - \omega)^2}\,, \tag{7.58}$$

which is identical to the nutation frequency Ω of (7.15) discussed in Sect. 7.1. In particular, at resonance, $\omega = \omega_0$, when the initial phase of the perturbation is zero, $\boldsymbol{\varrho}$ rotates in the $Y'Z'$ plane with angular velocity $\Omega = -|\mu_{12}\mathscr{E}|\hbar$.

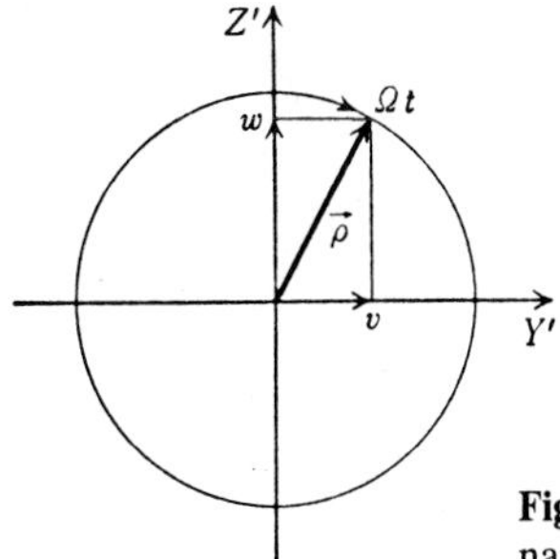

Fig. 7.6. Motion of a Bloch vector due to perturbation at the resonance frequency ($\omega = \omega_0$)

As seen from Fig. 7.6, the population inversion varies sinusoidally and the induced dipole moment is modulated with a 90° phase lag from the population inversion. When the initial phase of the perturbation is zero, $|\mu_{12}\mathscr{E}|$ is along the X' direction ($\boldsymbol{F}-\boldsymbol{\omega}$ is in the $-X'$ direction), and the dipole moment $\boldsymbol{p}$ given by $\varrho_{Y'}$, is 90° ahead in phase with respect to the complex vector $\mathscr{E}$ rotating with angular velocity $+\omega$, viewed in the stationary frame. This is a diagrammatic indication that the induced dipole moment of a two-level atom with population inversion is expressed by (7.21).

Let us now obtain the equations of motion of $\boldsymbol{\varrho}$ in the rotating frame in terms of its components. Denote the X', Y', and Z' components of $\boldsymbol{\varrho}$ by u, v, and w, respectively[3]. Since the X' and Y' axes are rotating with angular velocity ω about the $Z = Z'$ axis with respect to the X and Y axes, we have

$$(u+\mathrm{i}v)\,\mathrm{e}^{\mathrm{i}\omega t} = \varrho_X + \mathrm{i}\varrho_Y = 2\varrho_{12}\,, \qquad w = \varrho_Z\,. \tag{7.59}$$

Differentiating this with respect to t, we have

$$\frac{du}{dt} - \omega v + \mathrm{i}\left(\frac{dv}{dt} + \omega u\right) = 2\,\mathrm{e}^{-\mathrm{i}\omega t}\frac{d}{dt}\varrho_{12}\,,$$

$$\frac{dw}{dt} = \frac{d}{dt}\varrho_Z\,.$$

Substituting (7.42) and putting $x \equiv \mu_{12}\mathscr{E}/\hbar$ as was done in Sect. 7.1, we obtain the result for the perturbation (7.54):

$$\frac{du}{dt} = -(\omega_0-\omega)\,v - \mathrm{Im}\,\{x\}\,w\,, \tag{7.60 a}$$

$$\frac{dv}{dt} = (\omega_0-\omega)\,u + \mathrm{Re}\,\{x\}\,w\,, \tag{7.60 b}$$

$$\frac{dw}{dt} = \mathrm{Im}\,\{x\}\,u - \mathrm{Re}\,\{x\}\,v\,. \tag{7.60 c}$$

Needless to say, these are equivalent to (7.56).

7.5.3 Terms Representing Longitudinal Relaxation and Transverse Relaxation

So far we have described the Bloch equations and their geometrical representation for cases with no relaxation. The ensuing effects of relaxation processes can be treated by adding phenomenological terms as given below. Denoting the longitudinal relaxation constant by $1/\tau$ and the transverse rela-

---

3 $\boldsymbol{r} = (r_1, r_2, r_3)$ is sometimes used instead of $\boldsymbol{\varrho} = (u, v, w)$.

xation constant by γ, as in Sect. 7.4, we add the relaxation terms to (7.60) and obtain

$$\frac{du}{dt} = -\gamma u - (\omega_0 - \omega)\, v - \mathrm{Im}\,\{x\}\, w\ , \tag{7.61 a}$$

$$\frac{dv}{dt} = (\omega_0 - \omega)\, u - \gamma v + \mathrm{Re}\,\{x\}\, w\ , \tag{7.61 b}$$

$$\frac{dw}{dt} = \mathrm{Im}\,\{x\}\, u - \mathrm{Re}\,\{x\}\, v - \frac{w - w^{(0)}}{\tau}\ . \tag{7.61 c}$$

Here, $w^{(0)} = \Gamma_{\mathrm{ex}}\tau$ is the population inversion in the absence of perturbation (the value of w at $t = \infty$ when $\mathscr{E} = 0$). Validity of using constant relaxation rates to describe transient effects and saturation broadening for impurity ions in solids has recently been tested critically [7.2, 3].

When the longitudinal and transverse relaxation times are not equal, it is in most cases more convenient to deal with the equations of motion in terms of the components of $\boldsymbol{\varrho}$, rather than in vectorial form. However, with $\boldsymbol{i}, \boldsymbol{j}$, and $\boldsymbol{k}$ as the respective unit vectors of the rotating frame (X', Y', Z') in the abstract space, (7.61 a–c) can be collectively written in the form

$$\frac{d\boldsymbol{\varrho}}{dt} = (\boldsymbol{F} - \boldsymbol{\omega}) \times \boldsymbol{\varrho} - \gamma u \boldsymbol{i} - \gamma v \boldsymbol{j} - \frac{w - w^{(0)}}{\tau} \boldsymbol{k}\ , \tag{7.62}$$

where the relaxation terms are added to the equation of motion (7.56).

It is clear from (7.61) or (7.62) that the longitudinal relaxation represents the relaxation in the Z direction, while the transverse relaxation represents the relaxation in directions perpendicular to the Z axis. If the longitudinal relaxation constant τ^{-1} is zero and the transverse relaxation time is finite, $\boldsymbol{\varrho}$ moves in the stationary frame (X, Y, Z) as shown in Fig. 7.7 a, while it moves as shown in Fig. 7.7 b when γ is zero and τ is finite.

Problems

7.1 Find the solution of (7.12) under the initial conditions $a_1 = \mathrm{e}^{\mathrm{i}\theta}$ and $a_2 = 0$ at $t = 0$.

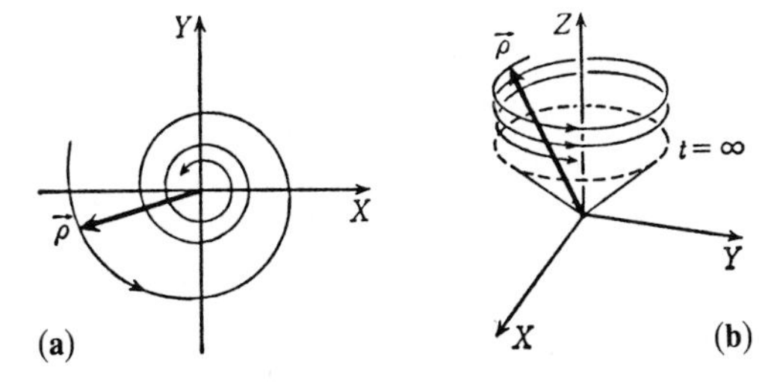

Fig. 7.7 a, b. Transverse relaxation (**a**) and longitudinal relaxation (**b**) of a Bloch vector

Answer: $a_1(t) = \exp[\mathrm{i}\theta + \mathrm{i}(\omega - \omega_0)t/2][\cos(\Omega t/2) - \mathrm{i}(\omega - \omega_0)\Omega^{-1}\sin(\Omega t/2)]$
$a_2(t) = \exp[\mathrm{i}\theta - \mathrm{i}(\omega - \omega_0)t/2](\mathrm{i}x^*/\Omega)\sin(\Omega t/2)$.

7.2 Using (7.21), calculate the phase of the induced dipole moment. Discuss how the phase changes with the amplitude of the optical field.

Answer: $\varphi = \tan^{-1}[\Omega(\omega - \omega_0)^{-1}\cot(\Omega t/2)]$, where Ω is a function of E.

7.3 Equation (7.1) can be integrated to give $\psi(t) = \exp(-\mathrm{i}\mathcal{H}_0 t/\hbar)\psi(0)$. When the unperturbed Hamiltonian is written as

$$\mathcal{H}_0 = \begin{pmatrix} W_1 & 0 \\ 0 & W_2 \end{pmatrix},$$

show that

$$\exp(-\mathrm{i}\mathcal{H}_0 t/\hbar) = \begin{pmatrix} \mathrm{e}^{-\mathrm{i}W_1 t/\hbar} & 0 \\ 0 & \mathrm{e}^{-\mathrm{i}W_2 t/\hbar} \end{pmatrix}.$$

7.4 Show that, for the ensemble-averaged density matrix $\langle\varrho\rangle_{\mathrm{av}}$, $\mathrm{Tr}(\langle\varrho\rangle_{\mathrm{av}}^2) \leqq 1$. When does the equality hold?

Answer: For the pure state.

7.5 Show that the ensemble average of the induced dipole moment is calculated from (7.20) to be (7.43).

7.6 Write explicitly the solutions of (7.51 a–c) in the absence of perturbation in terms of their initial values at $t = 0$.

Answer: $\varrho_X(t) = \varrho_X(0)\cos\omega_0 t - \varrho_Y(0)\sin\omega_0 t$,
$\varrho_Y(t) = \varrho_X(0)\sin\omega_0 t + \varrho_Y(0)\cos\omega_0 t$,
$\varrho_Z(t) = \varrho_Z(0)$.

7.7 Find the time dependence of the magnitude of the optical Bloch vector $\boldsymbol{\varrho}$ in the absence of all perturbation except relaxation effects.

Answer: $|\varrho|^2 = (w^{(0)})^2 + 2w^{(0)}(w_0 - w^{(0)})\mathrm{e}^{-t/\tau}$
$+ (w_0 - w^{(0)})^2\mathrm{e}^{-2t/\tau} + (u_0^2 + v_0^2)\mathrm{e}^{-2\gamma t}$,
where u_0, v_0, and w_0 are initial values of u, v, and w, respectively.

7.8 Solve (7.60 a–c) to obtain the components of $\boldsymbol{\varrho}$ in the rotating frame, u, v, and w, when the perturbation field is $E(t) = E_0\cos\omega t$ and μ_{12} is real.

Hint: $\mathrm{Im}\{x\} = 0$.
Answer: $u = (1 - \cos\Omega t)x(\omega - \omega_0)/\Omega^2$
$v = (\sin\Omega t)x/\Omega$
$w = [(\omega - \omega_0)^2 + x^2\cos\Omega t]/\Omega^2$.

8. Nonlinear Coherent Effects

In ordinary optical phenomena, such quantities as the absorption coefficient and the refractive index of a medium are independent of the light intensity and are constant. But, as may be understood from the discussion in the preceding chapter, the effects of a coherent interaction between the laser field and matter are nonlinear with intensity and not stationary in time. For a weak field, the polarization $\boldsymbol{P}$ of the medium is proportional to the optical field $\boldsymbol{E}$ and is linear, but it generally becomes nonlinear for a strong field. When the polarization $\boldsymbol{P}$ is not proportional to the field $\boldsymbol{E}$, the susceptibility χ is not constant, nor are the absorption coefficient and the refractive index. As a result, nonlinear optical phenomena appear, which could not be expected in ordinary linear optics.

The stronger the light intensity, the more prominent are the nonlinear optical effects. However, the appearance of nonlinear effects is not simply determined by the magnitude of the optical power or energy. With ordinary light hardly any nonlinear optical effect is observed even with a power of higher than 100 W, whereas remarkable nonlinear optical effects can be observed even with a laser power of below 1 mW. In a short laser pulse the resonant coherent interaction can be very nonlinear even with an energy of less than 1 nJ.

The nonlinear optical effects include saturated absorption, optical harmonic generation, optical frequency mixing, and a number of other phenomena. Here, we shall be mainly concerned with the basic nonlinear effects of the coherent interaction.

8.1 Saturation Effect

In Sect. 4.6 the complex susceptibility and the absorption constant for weak incident light were obtained. The coherent interaction between a two-level atom (or molecule) and a monochromatic field of arbitrary amplitude was calculated in Chap. 7. The induced dipole moment given by (7.21) is nonlinear, because Ω involves x, which is proportional to the amplitude of the incident light. In this chapter we shall obtain the nonlinear absorption coefficient of a homogeneous medium consisting of two-level atoms, using the results of the preceding chapter.

Suppose that some of the many two-level atoms in a medium are in the upper level and others are in the lower level. When a coherent field as given by (7.4) is incident, the atoms in the upper level will make induced emission and those in the lower level will absorb the incident light. The probabilities of atoms in either level undergoing transitions induced by the incident field during a time from 0 to t are given by (7.18). If the relaxation effect is taken into consideration, the duration of the coherent interaction between the atoms and the field will be finite even though the incident field is a steady wave of infinite duration.

For example, a molecule in a gas immediately after a collision with another molecule must be in either the upper or the lower level; it will thereafter be perturbed by the coherent interaction with the incident field, but the interaction will be terminated at the instant when the molecule subsequently collides with any other molecule. The collision between molecules is random, so that the probability that a collision takes place between times t and $t + dt$ may be denoted by a constant, $\gamma\, dt$. Then, the probability that the coherent interaction should have a duration of between t and $t + dt$ is given by

$$w(t)\, dt = \gamma \mathrm{e}^{-\gamma t}\, dt \,. \tag{8.1}$$

Now suppose that each molecule is strongly perturbed during the short collision so that its state afterwards, which we assume is one of the pure molecular eigenstates, is independent of its state before. Then the average duration of the coherent interaction is equal to the relaxation time $\tau = 1/\gamma$ for longitudinal and transverse relaxations.

Therefore, the average probability that an atom (or a molecule), which was in the upper level immediately after a collision, should make a transition to the lower level before the next collision, is given by

$$\langle |a_1(t)|^2 \rangle_{\mathrm{av}} = \int_0^\infty w(t) |a_1(t)|^2\, dt \,.$$

Substituting (7.18) and (8.1), this becomes

$$\langle |a_1(t)|^2 \rangle_{\mathrm{av}} = \frac{|x|^2}{2(\Omega^2 + \gamma^2)} \,. \tag{8.2}$$

When this is divided by the average duration of the coherent interaction $\tau = 1/\gamma$, it gives the transition probability $p(2 \to 1)$ per unit time. The transition probability $p(1 \to 2)$ per unit time that an atom in the lower level makes a transition to the upper level by absorbing the incident light is the same and is given by

$$p(1 \to 2) = p(2 \to 1) = \frac{\gamma |x|^2}{2(\Omega^2 + \gamma^2)} \,. \tag{8.3}$$

Now, suppose that the incident light is absorbed by a medium in which the number of atoms in the lower level per unit volume is $N_1^{(0)}$, and in the upper level is $N_2^{(0)}$. Then the power absorbed by atoms in the lower level, i.e., the energy absorbed per unit time, is $N_1^{(0)}\hbar\omega p\,(1 \to 2)$, and the power of induced emission from atoms in the upper level is $N_2^{(0)}\hbar\omega p\,(2 \to 1)$. Using (8.3), we obtain the net power absorbed by a unit volume of the medium:

$$\Delta P = (N_1^{(0)} - N_2^{(0)})\,\frac{\hbar\omega\gamma|x|^2}{2(\Omega^2 + \gamma^2)}\,. \tag{8.4}$$

Since the amplitude of the incident light field is $\mathscr{E}$, as shown by (7.4), the power density (the incident light power per unit cross-section) is given by[1]

$$P = \frac{\varepsilon_0|\mathscr{E}|^2}{2}c\,. \tag{8.5}$$

The power absorbed per unit volume is written

$$\Delta P = 2\alpha P \tag{8.6}$$

where α is the amplitude absorption coefficient. Therefore, using (8.4–6) and $\Omega^2 = (\omega - \omega_0)^2 + |x|^2$, we obtain

$$\alpha = \frac{(N_1^{(0)} - N_2^{(0)})\,\omega\gamma|\mu_{12}|^2}{2\,\varepsilon_0\hbar c\,[(\omega_0 - \omega)^2 + \gamma^2 + |x|^2]}\,. \tag{8.7}$$

In most cases in the visible, $N_2^{(0)} \ll N_1^{(0)}$, but there are cases where $N_2^{(0)}$ is not small.

Here, we have assumed that the resonance frequency of the two-level atoms of the medium is homogeneous and equal to ω_0. Figure 8.1 shows the

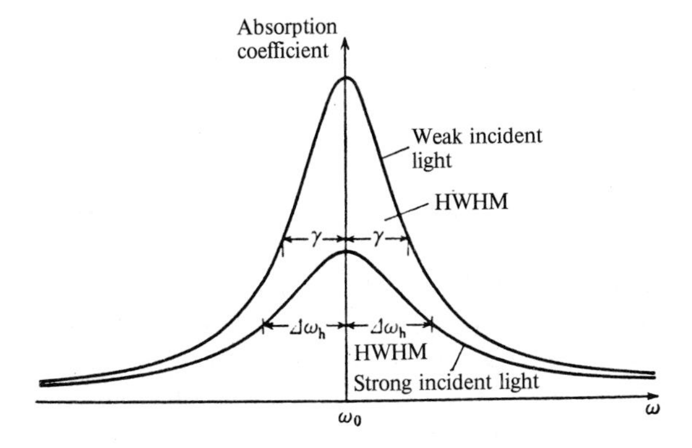

Fig. 8.1. Profile of a homogeneously broadened spectral line

1 The electric energy density is $(1/2)\,\varepsilon_0|E(t)|^2 = (1/4)\,\varepsilon_0|\mathscr{E}|^2$, and the magnetic energy density has the same magnitude; hence their sum is $(1/2)\,\varepsilon_0|\mathscr{E}|^2$.

absorption coefficient (8.7) plotted as a function of frequency for different incident light intensities. The profile of the homogeneous line is Lorentzian, whether the incident light is strong or weak. its halfwidth at half maximum is called the homogeneous width as opposed to the inhomogeneous width described in the next section. It is obtained from (8.7):

$$\Delta\omega_h = \sqrt{\gamma^2 + |x|^2} \, . \tag{8.8}$$

The linewidth is seen to be broadened with increasing incident light intensity, as shown in Fig. 8.2. This is called saturation broadening.

Again, as the light intensity is increased at a constant frequency, the absorbed power increases linearly while the light intensity is weak. As the light becomes stronger, the absorption coefficient becomes gradually smaller, and the absorbed power approaches a certain saturation value denoted by ΔP_s, as shown in Fig. 8.3. Since the absorbed power is no longer proportional to the incident power, such an effect is called saturated absorption or nonlinear absorption.

The amplitude absorption coefficient is expressed as a function of the power density P of the incident light in the form

$$\alpha(P) = \frac{\alpha(0)}{1 + P/P_s} \, . \tag{8.9}$$

Here $\alpha(0)$ is the amplitude absorption coefficient for $P = 0$, i.e., the linear absorption constant, which, at the central frequency $\omega = \omega_0$, is

$$\alpha(0) = (N_1^{(0)} - N_2^{(0)}) \frac{\omega|\mu_{12}|^2}{2\varepsilon_0 \hbar c \gamma} \, . \tag{8.10}$$

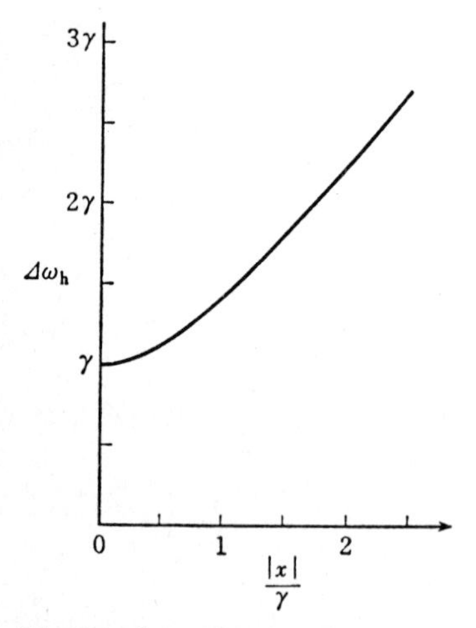

Fig. 8.2. Saturation broadening of the homogeneous width. $\Delta\omega_h$ is the halfwidth at half maximum, and $x = \mu_{12}\mathscr{E}/\hbar$

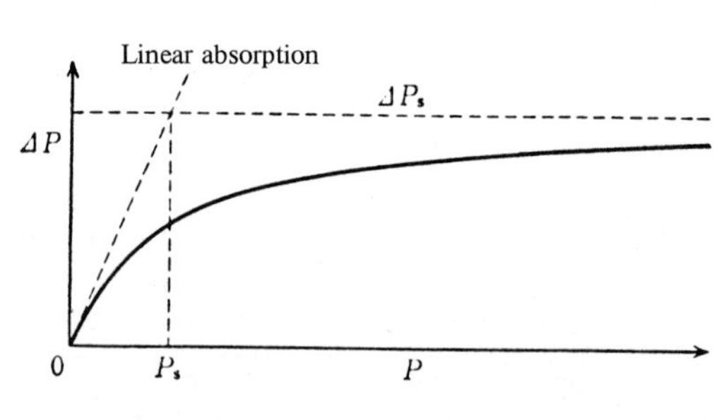

Fig. 8.3. Saturation of the absorbed power as a function of the incident power. P_s is the saturation power, and ΔP_s is the saturation value of the absorbed power

This is the same as (4.42) where (4.43) has been substituted for $g(\omega)$, with $\Delta\omega = \gamma$ and putting $|\mu_{UL}|^2/3 = |\mu_{12}|^2$. The P_s used in (8.9) is called the saturation power[2] or the saturation parameter, and is given by

$$P_s = \frac{\varepsilon_0 \hbar^2 c \gamma^2}{2|\mu_{12}|^2} . \tag{8.11}$$

The saturation value ΔP_s of the absorbed power is expressed from (8.9) and (8.6), for $P \to \infty$, as $\Delta P_s = 2\alpha(0) P_s$. Substituting (8.11) we obtain $\Delta P_s = \frac{1}{2}(N_1^{(0)} - N_2^{(0)}) \hbar\omega\gamma$.

8.2 Change in the Atomic Populations Due to Saturated Absorption

Equation (8.7) for the absorption coefficient is expressed in such a form as to be proportional to the unperturbed population difference $N_1^{(0)} - N_2^{(0)}$. In saturated absorption, however, N_1 becomes less than $N_1^{(0)}$ with a corresponding increase in N_2 over $N_2^{(0)}$. The average transition probability (8.2) is used to calculate the perturbed populations N_1 and N_2, with the result

$$\begin{aligned} N_1 &= N_1^{(0)} - (N_1^{(0)} - N_2^{(0)}) \frac{|x|^2}{2(\Omega^2 + \gamma^2)} , \\ N_2 &= N_2^{(0)} + (N_1^{(0)} - N_2^{(0)}) \frac{|x|^2}{2(\Omega^2 + \gamma^2)} . \end{aligned} \tag{8.12}$$

Then the atomic population difference becomes

$$\begin{aligned} N_1 - N_2 &= (N_1^{(0)} - N_2^{(0)}) \left(1 - \frac{|x|^2}{\Omega^2 + \gamma^2}\right) \\ &= (N_1^{(0)} - N_2^{(0)}) \frac{(\omega_0 - \omega)^2 + \gamma^2}{(\omega_0 - \omega)^2 + \gamma^2 + |x|^2} . \end{aligned} \tag{8.13}$$

Rather than saying that this is the result of saturated absorption, it should be considered as being its cause, because the decrease in the absorption coefficient for strong incident light is due to the decrease in the population difference.

The above statement can be understood by noting that the power absorption coefficient 2α can be written

$$2\alpha = (N_1 - N_2)\sigma ,$$

2 More rigorously it is the saturation power density.

as shown by (4.45), where σ is the absorption cross-section. As the light intensity increases, σ is constant so that the decrease in α can be explained by the decrease in $N_1 - N_2$, as shown below. The absorption cross-section corresponding to saturated absorption (8.7) is obtained from (8.13) to be

$$\sigma = \frac{\omega\gamma|\mu_{12}|^2}{\varepsilon_0\hbar c\,[(\omega_0 - \omega)^2 + \gamma^2]}\,, \tag{8.14}$$

which is independent of $|x|^2$.

So far we have considered the absorption of ideal two-level atoms by assuming the longitudinal and the transverse relaxation times to be equal. In actual light absorbers, however, there exist the ground level and levels other than the upper and the lower levels (2 and 1) of the transition. Thus the atoms excited to levels 1 or 2 may make transitions statistically to any one of those levels, and, since their transition probabilities are different, the quantity $N_1 + N_2$ cannot be conserved. Even in such a case, the rate-equation approximation is often used by treating the relaxation phenomenologically, following the treatment of two-level atoms but using different relaxation rates γ_1 and γ_2 for the level 1 and 2. The rate equations for such quasi-two-level atoms may be written as

$$\begin{aligned} \frac{dN_1}{dt} &= -\gamma_1(N_1 - N_1^{(0)}) - (N_1 - N_2)\,I\sigma\,, \\ \frac{dN_2}{dt} &= -\gamma_2(N_2 - N_2^{(0)}) + (N_1 - N_2)\,I\sigma\,, \end{aligned} \tag{8.15}$$

where I is the flux of incident photons and the absorption cross-section σ is that given by replacing γ in (8.14)[3]:

$$\gamma = \tfrac{1}{2}(\gamma_1 + \gamma_2)\,. \tag{8.16}$$

In terms of the power density P of the incident light, or $|x|^2$, I is expressed as

$$I = \frac{P}{\hbar\omega} = \frac{\varepsilon_0\hbar c}{2\omega|\mu_{12}|^2}\,|x|^2\,. \tag{8.17}$$

3 This follows from the feasibility of considering the indeterminacy of the energy of the upper and lower levels to be $h\gamma_2/2$ and $h\gamma_1/2$, respectively, and taking the linewidth of the transition between the two levels to be the sum of $\gamma_2/2$ and $\gamma_1/2$. For the Lorentzian function we have

$$\frac{1}{\pi^2}\int_{-\infty}^{\infty}\frac{a}{(x-y)^2 + \alpha^2}\cdot\frac{b}{(y-x_0)^2 + b^2}\,dy = \frac{1}{\pi}\cdot\frac{a+b}{(x-x_0)^2 + (a+b)^2}$$

so that the convolution of the two widths a and b gives $a + b$. However, such a simple sum is not valid for other types of lineshape.

Now, since (8.15) is zero in the steady state, when the incident light intensity is constant in the medium we obtain

$$N_1 = N_1^{(0)} - \frac{\gamma_2 (N_1^{(0)} - N_2^{(0)})\, I\sigma}{\gamma_1\gamma_2 + (\gamma_1 + \gamma_2)\, I\sigma},$$

$$N_2 = N_2^{(0)} + \frac{\gamma_1 (N_1^{(0)} - N_2^{(0)})\, I\sigma}{\gamma_1\gamma_2 + (\gamma_1 + \gamma_2)\, I\sigma}. \tag{8.18}$$

From (8.14, 17), we have

$$I\sigma = \frac{|x|^2}{2} \cdot \frac{\gamma}{(\omega_0 - \omega)^2 + \gamma^2}. \tag{8.19}$$

Using the longitudinal relaxation time defined by

$$\tau = \frac{1}{2}\left(\frac{1}{\gamma_1} + \frac{1}{\gamma_2}\right), \tag{8.20}$$

we may rewrite (8.18) to obtain

$$N_1 = N_1^{(0)} - \frac{\gamma}{2\gamma_1} \cdot \frac{(N_1^{(0)} - N_2^{(0)})|x|^2}{(\omega_0 - \omega)^2 + \gamma^2 + \gamma\tau|x|^2},$$

$$N_2 = N_2^{(0)} + \frac{\gamma}{2\gamma_2} \cdot \frac{(N_1^{(0)} - N_2^{(0)})|x|^2}{(\omega_0 - \omega)^2 + \gamma^2 + \gamma\tau|x|^2}. \tag{8.21}$$

Therefore, the population difference is expressed as

$$N_1 - N_2 = (N_1^{(0)} - N_2^{(0)}) \left(1 - \frac{\gamma\tau|x|^2}{(\omega_0 - \omega)^2 + \gamma^2 + \gamma\tau|x|^2}\right). \tag{8.22}$$

When $\gamma_1 = \gamma_2$, $\gamma\tau = 1$ and this equation is reduced to (8.13). When γ_1 and γ_2 are different, as in quasi-two-level atoms, the decrease in N_1 is not equal to the increase in N_2.

The power absorption coefficient $2\alpha = (N_1 - N_2)\sigma$ is then obtained from (8.14, 22) to be

$$2\alpha = \frac{N_1^{(0)} - N_2^{(0)}}{\varepsilon_0 \hbar c} \cdot \frac{\omega\gamma|\mu_{12}|^2}{(\omega_0 - \omega)^2 + \gamma^2 + \gamma\tau|x|^2}. \tag{8.23}$$

This is consistent with the previous calculation (8.7) when $\gamma\tau = 1$. When $\gamma\tau \neq 1$, the expression for the saturation power becomes

$$P_s = \frac{\varepsilon_0 \hbar^2 c \gamma}{2|\mu_{12}|^2 \tau} \tag{8.24}$$

instead of (8.11). Then the saturation photon flux $I_s = P_s/\hbar\omega$ is given by

$$I_s = \frac{1}{2\sigma\tau}, \tag{8.25}$$

and the relation (8.9) between the saturated absorption coefficient $\alpha(I)$ and the linear absorption constant $\alpha(0)$ can be expressed as

$$\alpha(I) = \frac{\alpha(0)}{1 + I/I_s}.$$

The gain constant of a medium with population inversion ΔN saturates in a similar manner. Thus, denoting the unsaturated gain constant for weak incident light by $G(0)$, the gain coefficient with incident power P or photon flux I can be written

$$G(P) = \frac{G(0)}{1 + P/P_s} = \frac{G(0)}{1 + I/I_s} \tag{8.26}$$

where P_s and I_s are the same as in (8.24, 25). $G(P)$ is called the saturated gain coefficient and $G(0)$ the small-signal gain constant.

8.3 Nonlinear Complex Susceptibility

Now, it is also possible to investigate the saturated absorption of two-level atoms by using the ensemble average of the density matrix. We shall thus obtain the complex susceptibility which describes not only the nonlinear absorption but also the nonlinear dispersion.

When we consider the steady-state absorption of light, we have

$$\begin{aligned} &\gamma u + (\omega_0 - \omega)v + \mathrm{Im}\{x\}w = 0, \\ &(\omega_0 - \omega)u - \gamma v + \mathrm{Re}\{x\}w = 0, \\ &\mathrm{Im}\{x\}u - \mathrm{Re}\{x\}v - (w - w^{(0)})/\tau = 0, \end{aligned} \tag{8.27}$$

since each of the equations (7.61) should be zero. Solving these simultaneous equations, we obtain

$$u = \frac{-(\omega_0 - \omega)\,\mathrm{Re}\{x\} - \gamma\,\mathrm{Im}\{x\}}{(\omega_0 - \omega)^2 + \gamma^2 + \gamma\tau|x|^2} w^{(0)}, \tag{8.28 a}$$

$$v = \frac{\gamma\,\mathrm{Re}\{x\} - (\omega_0 - \omega)\,\mathrm{Im}\{x\}}{(\omega_0 - \omega)^2 + \gamma^2 + \gamma\tau|x|^2} w^{(0)}, \tag{8.28 b}$$

$$w = \frac{(\omega_0 - \omega)^2 + \gamma^2}{(\omega_0 - \omega)^2 + \gamma^2 + \gamma\tau|x|^2} w^{(0)}. \tag{8.28 c}$$

The induced dipole moment of the two-level atom is given by (7.43), which becomes, substituting (7.59),

$$\langle p(t)\rangle_{\mathrm{av}} = \tfrac{1}{2}(u + \mathrm{i}v)\,\mathrm{e}^{\mathrm{i}\omega t}\mu_{21} + \mathrm{c.c.}$$

Then the macroscopic polarization of the medium is given by

$$P(t) = \tfrac{1}{2} N\mu_{21}(u + \mathrm{i}v)\,\mathrm{e}^{\mathrm{i}\omega t} + \mathrm{c.c.} \tag{8.29}$$

where N is the number of two-level atoms in a unit volume of the medium. By using

$$u + \mathrm{i}v = \frac{(-\omega_0 + \omega + \mathrm{i}\gamma)x}{(\omega_0 - \omega)^2 + \gamma^2 + \gamma\tau|x|^2}\,w^{(0)}$$

from (8.28 a, b) and $x = \mu_{12}\,\mathscr{E}/\hbar$, (8.29) can be expressed as

$$P(t) = \frac{Nw^{(0)}}{2\hbar}\cdot\frac{-\omega_0 + \omega + \mathrm{i}\gamma}{(\omega_0 - \omega)^2 + \gamma^2 + \gamma\tau|x|^2}\,|\mu_{12}|^2\,\mathscr{E}\,\mathrm{e}^{\mathrm{i}\omega t} + \mathrm{c.c.} \tag{8.30}$$

where $|\mu_{12}|^2 = \mu_{21}\mu_{12}$.

Since the electric field of the incident light (7.4) is

$$E(t) = \tfrac{1}{2}\,\mathscr{E}\,\mathrm{e}^{\mathrm{i}\omega t} + \mathrm{c.c.}\,,$$

the nonlinear susceptibility of saturated absorption is expressed as

$$\chi = \chi' - \mathrm{i}\chi'' = \frac{N_1^{(0)} - N_2^{(0)}}{\varepsilon_0\hbar}|\mu_{12}|^2\frac{\omega_0 - \omega - \mathrm{i}\gamma}{(\omega_0 - \omega)^2 + \gamma^2 + \gamma\tau|x|^2}\,. \tag{8.31}$$

Here, we have put

$$Nw^{(0)} = -(N_1^{(0)} - N_2^{(0)})\,,$$

because the population inversion $w = \varrho_Z = \varrho_{22} - \varrho_{11}$ for an absorber is negative.

When $|\chi'| \ll 1$ in the complex susceptibility, the power absorption coefficient is given by

$$2\alpha = \frac{\omega}{c}\chi''\,. \tag{8.32}$$

Since the imaginary part of (8.31) is $-\chi''$, its substitution into (8.32) gives a result identically equal to (8.23).

Moreover, the real part of (8.31) represents the nonlinear dispersion, given by

$$\chi' = \frac{N_1^{(0)} - N_2^{(0)}}{\varepsilon_0 \hbar} |\mu_{12}|^2 \frac{\omega_0 - \omega}{(\omega_0 - \omega)^2 + \gamma^2 + \gamma\tau |x|^2} . \tag{8.33}$$

When $|\chi'| \ll 1$, the nonlinear refractive index becomes

$$\eta \simeq \sqrt{1 + \chi'} \simeq 1 + \tfrac{1}{2} \chi' .$$

8.4 Inhomogeneous Broadening

In the preceding section we assumed that the resonance frequencies of two-level atoms in a medium are all equal; in an actual medium, however, the resonance frequencies differ slightly depending on the position and orientation of the atoms. Moreover, the velocities of the molecules in a gas vary from one molecule to another and their resonance frequencies are spread owing to the Doppler effect. When the resonance frequencies of atoms or molecules in a medium are not homogeneous, the spectral line of the medium is broadened accordingly. This is called inhomogeneous broadening [8.1]. It is not possible, however, to discuss the inhomogeneous broadening in common due to causes such as the distribution of resonance frequencies of atoms in a crystal, and the inhomogeneity of the composition of the medium, or that of the external magnetic field, and so forth.

8.4.1 Doppler Broadening

Since the distribution of molecular velocities in a gas is of the Maxwell-Boltzmann type, the spectral line is broadened by the Doppler effect to take a Gaussian line shape. This is called the Doppler broadening [8.1].

The Doppler effect due to the molecular motion is mainly due to the velocity component along the direction of the incident light. The second-order Doppler effect is independent of the direction of the molecular velocity with respect to the incident light, but its relative magnitude is of the order of 10^{-12}, so that it may be neglected, except for super-high precision. Now, denoting the molecular velocity component along the direction of light propagation by v, the frequency of the incident light as seen by the molecule shifts from ω to ω', due to the Doppler effect, where ω' is

$$\omega' = \omega - kv , \tag{8.34}$$

and $k = \omega/c$. When the difference of eigenenergies of the molecule is $\hbar\omega_0 = W_2 - W_1$, resonance does not occur at $\omega = \omega_0$ but at $\omega' = \omega_0$. That is to say, resonance occurs when the frequency of the incident light is

$$\omega = \omega_0 + kv \ . \tag{8.35}$$

Since the molecular velocity component v is distributed in both positive and negative directions, the spectral line is inhomogeneously broadened in accordance with the velocity distribution.

Now, assuming the Maxwell-Boltzmann velocity distribution, the number of molecules in a gas with velocity components between v and $v + dv$ is given by

$$N(v)\,dv = \frac{N}{\sqrt{\pi}u} \mathrm{e}^{-v^2/u^2}\,dv \ . \tag{8.36}$$

Here, u is the most probable velocity, which is given by

$$u = \sqrt{\frac{2k_{\mathrm{B}}T}{M}} \ , \tag{8.37}$$

where T is the temperature of the gas, M the mass of the molecule and k_{B} the Boltzmann constant.

By substituting the velocity component $v = (\omega - \omega_0)/k$, obtained from (8.35), into (8.36), the spectral distribution of the resonance frequency becomes

$$g(\omega) = \frac{1}{\sqrt{\pi}\,ku} \exp\left[-\left(\frac{\omega - \omega_0}{ku}\right)^2\right] . \tag{8.38}$$

This is a Gaussian profile (Sect. 4.5) and its halfwidth at half maximum is

$$\Delta\omega_{\mathrm{D}} = \sqrt{\ln 2}\,ku = 0.833\,ku \ , \tag{8.39}$$

which is called the Doppler width (or Doppler halfwidth). Even when the cause of the inhomogeneous broadening is not the Doppler effect, as in the spectral line of a solid, the calculation given below can be approximately applied, provided the inhomogeneously broadened profile of the spectral line is approximately Gaussian.

If the homogeneous broadening in the spectral line of a gas is small, its lineshape is almost entirely determined by the Doppler broadening, which is a Gaussian profile with a Doppler width. However, in order to observe the spectral line in practice the transition probability must not be too small, and there must be a fairly large number of molecules so that collisions between the molecules take place. Thus, the relaxation time is finite and the homogeneous broadening does not vanish. The homogeneous broadening due to the natural lifetime of the levels is called the natural width, and that due to molecular collisions is called the collisional width. Since the collisional width is almost proportional to the gas pressure, it is sometimes called pressure broadening. Saturation broadening is also homogeneous.

In general, the homogeneous broadening occurs simultaneously with the Doppler broadening, and the resulting spectral line is broadened, as shown in Fig. 8.4. Its halfwidth at half maximum (HWHM) is broader than either the total homogeneous width $\Delta\omega_h$ or the Doppler width $\Delta\omega_D$ alone[4]. Yet, the width is not equal to the sum $\Delta\omega_h + \Delta\omega_D$ of the two, but, as shown in Fig. 8.4, it is somewhat narrower.

8.4.2 Nonlinear Susceptibility with Doppler Broadening

The profile of a spectral line with both homogeneous and inhomogeneous broadening is expressed by a convolution integral of the two broadenings. In order to obtain the line shape of not only the absorption or the emission, but also the dispersion, let us calculate the nonlinear complex susceptibility, taking into account the Doppler effect [8.2]. Since the resonance frequency ω_0 in (8.31) with a Doppler effect becomes $\omega_0 + kv$, and the distribution of the velocity component v is expressed by (8.36), the nonlinear complex susceptibility of a Doppler broadened line is given by

$$\chi_D = \frac{N_1^{(0)} - N_2^{(0)}}{\sqrt{\pi}\,\varepsilon_0 \hbar u} |\mu_{12}|^2 \int_{-\infty}^{\infty} \frac{(\omega_0 + kv - \omega - i\gamma)\, e^{-v^2/u^2}\, dv}{(\omega_0 + kv - \omega)^2 + \gamma^2 + \gamma\tau|x|^2} \,. \tag{8.40}$$

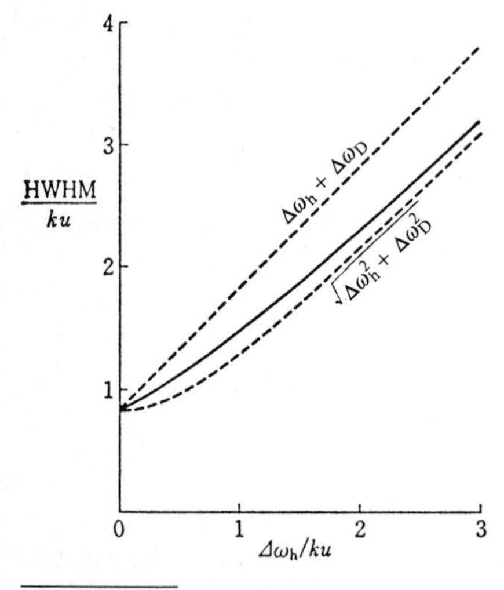

Fig. 8.4. Halfwidth at half maximum of a spectral line with both Doppler broadening and homogeneous broadening (*solid line*). $\Delta\omega_h$ is the homogeneous width and $\Delta\omega_D = 0.833\,ku$ is the Doppler width

---

4 If the phase of light emitted or absorbed by a molecule does not change appreciably on collision with another molecule, the width of the spectral line can be narrower than the Doppler broadening. This extraordinary effect is called Dicke narrowing. On the contrary, in the usual collisions between molecules, phase-changing collisions are more frequent than energy-changing collisions. Therefore, even if the longitudinal relaxation time is long, the transverse relaxation time can be shorter. Calculations in the previous section may be applied to such cases by considering γ and τ to be phenomenological parameters, different from (8.16 and 20), respectively.

The real part χ'_D of χ_D represents the dispersion, and the imaginary part, taken as $-\chi''_D$, gives the power absorption coefficient $k\chi''_D$, from (8.32). Accordingly, the power absorption coefficient 2α, with Doppler and homogeneous broadening, is expressed as a function of the frequency and amplitude of the incident light:

$$2\alpha_D = \frac{N_1^{(0)} - N_2^{(0)}}{\sqrt{\pi}\,\varepsilon_0\hbar u}|\mu_{12}|^2 k\gamma \int_{-\infty}^{\infty} \frac{e^{-v^2/u^2}\,dv}{(\omega_0 + kv - \omega)^2 + \gamma^2 + \gamma\tau|x|^2}\,. \tag{8.41}$$

The integrals appearing in these equations can be expressed in terms of plasma dispersion functions. With ζ as a complex variable, the plasma dispersion function $Z(\zeta)$ is defined as

$$Z(\zeta) = \frac{1}{\sqrt{\pi}} \int_{-\infty}^{\infty} \frac{e^{-x^2}\,dx}{x - \zeta}\,. \tag{8.42}$$

It may also be written as

$$Z(\zeta) = i\int_0^{\infty} \exp\left(i\zeta y - \frac{1}{4}y^2\right) dy\,, \quad \text{or}$$

$$Z(\zeta) = 2i\,e^{-\zeta^2} \int_{-\infty}^{i\zeta} e^{-t^2}\,dt\,,$$

where x and y are real variables and t is a complex variable.

Now, using the homogeneous width of saturation broadening,

$$\Delta\omega_h = \sqrt{\gamma^2 + \gamma\tau|x|^2}\,, \tag{8.43}$$

and putting

$$\zeta = \frac{\omega - \omega_0 - i\Delta\omega_h}{ku}\,, \qquad x = \frac{v}{u}\,,$$

we have

$$\frac{e^{-x^2}\,dx}{x - \zeta} = \frac{kv - \omega + \omega_0 - i\Delta\omega_h}{(kv - \omega + \omega_0)^2 + (\Delta\omega_h)^2}\,e^{-v^2/u^2}\,k\,dv\,,$$

so that we can obtain from (8.40)

$$\chi'_D = \frac{N_1^{(0)} - N_2^{(0)}}{\varepsilon_0\hbar ku}|\mu_{12}|^2\,\mathrm{Re}\{Z(\zeta)\}\,, \tag{8.44}$$

$$-\chi''_D = \frac{N_1^{(0)} - N_2^{(0)}}{\varepsilon_0\hbar ku}|\mu_{12}|^2 \frac{\gamma}{\Delta\omega_h}\,\mathrm{Im}\{Z(\zeta)\}\,. \tag{8.45}$$

In general, these express the dispersion and absorption in saturated absorption, whereas the linear susceptibility $\chi_1 = \chi_1' - i\chi_1''$, when the light is weak, is expressed by the limit of χ_D as $|x| \to 0$. Therefore, we have

$$\chi_1 = \frac{N_1^{(0)} - N_2^{(0)}}{\varepsilon_0 \hbar k u} |\mu_{12}|^2 Z\left(\frac{\omega - \omega_0 - i\gamma}{ku}\right) . \tag{8.46}$$

The polarization of a medium may, in general, be written in the form

$$P = \varepsilon_0 (\chi_1 E + \chi_2 E^2 + \ldots) \tag{8.47}$$

where E is the optical electric field and χ_n is the nth-order nonlinear susceptibility. When the medium is symmetric with respect to inversion of coordinates, the nonlinear susceptibilities of even orders will vanish. Therefore, for isotropic gases, liquids, and most solids, the nonlinear susceptibility is expressed as

$$\chi = \chi_1 + \chi_3 E^2 + \chi_5 E^4 + \ldots .$$

The odd-order nonlinear coefficients for the nonlinear susceptibility including the Doppler broadening (8.40) can be expressed in terms of $Z(\zeta)$ and its derivatives.

Whereas the Doppler broadening of a gas is as much as 10^9 Hz = 1 GHz in the visible range of the spectrum, the homogeneous broadening is only about 10^6–10^7 Hz = 1–10 MHz at pressures below 100 Pa.[5] We can then assume $\Delta\omega_h \ll \Delta\omega_D$, namely, $\gamma \ll ku$. The approximation with $\gamma \ll ku$ is called the Doppler-limit approximation, in which the plasma dispersion function takes the simple form

$$\mathrm{Re}\{Z(\zeta)\} = 2e^{-z^2} \int_0^z e^{x^2} dx , \quad \mathrm{Im}\{Z(\zeta)\} = -\sqrt{\pi} e^{-z^2} , \tag{8.48}$$

where $z = (\omega - \omega_0)/ku$. Thus the power absorption coefficient (8.45) in the Doppler-limit approximation (8.48) becomes

$$2\alpha_D = \frac{N_1^{(0)} - N_2^{(0)}}{\varepsilon_0 \hbar u} |\mu_{12}|^2 \frac{\sqrt{\pi}\gamma}{\Delta\omega_h} \exp\left[-\left(\frac{\omega - \omega_0}{ku}\right)^2\right] . \tag{8.49}$$

If we compare the absorption coefficient $\alpha_D(0)$ at $\omega = \omega_0$ for weak incident light with (8.10) for the case of no Doppler broadening, we find

$$\frac{\alpha_D(0)}{\alpha(0)} = \frac{\sqrt{\pi}\gamma}{ku} . \tag{8.50}$$

---

5 Approximately below 0.75 Torr.

This shows that the peak intensity of a spectral line is reduced by Doppler broadening by approximately the ratio of the homogeneous width to the Doppler width.

The saturation effect at the central frequency of an absorption line with Doppler broadening, i.e., the decrease in the absorption coefficient with the incident power P, is expressed in the Doppler-limit approximation as

$$\alpha_D(P) = \frac{\alpha_D(0)}{\sqrt{1 + P/P_s}}, \tag{8.51}$$

where P_s is the saturation power. Although P_s is the same as in the absence of Doppler broadening (8.24), the factor in (8.51) is $(1 + P/P_s)^{-1/2}$ in contrast to $(1 + P/P_s)^{-1}$ in (8.9) for the case of no Doppler broadening. Here (8.51) holds for $\gamma \ll ku$, while (8.9) holds for $\gamma \gg ku$.

If the Doppler-limit approximation does not apply, the integral in the plasma dispersion function must be calculated numerically. Figure 8.5 shows the numerically calculated ratio of the absorption coefficient (8.41) for weak light ($|x|^2 \to 0$) at the central frequency to that for the case with no Doppler broadening. It can be seen how the linear absorption coefficient $\alpha_D(0)$ for the case with Doppler broadening approaches the Doppler limit as the Doppler broadening increases, and how it approaches the value $\alpha(0)$ for the case with homogeneous broadening only, as the Doppler broadening decreases.

8.5 Hole-Burning Effect

The distinction between homogeneous and inhomogeneous broadening was first disclosed by the hole-burning effect in a nuclear magnetic resonance experiment. When a spectral line is broadened inhomogeneously as well as homogeneously, the saturation, in general, occurs only within the homogeneous width around the frequency of the incident light and does not extend

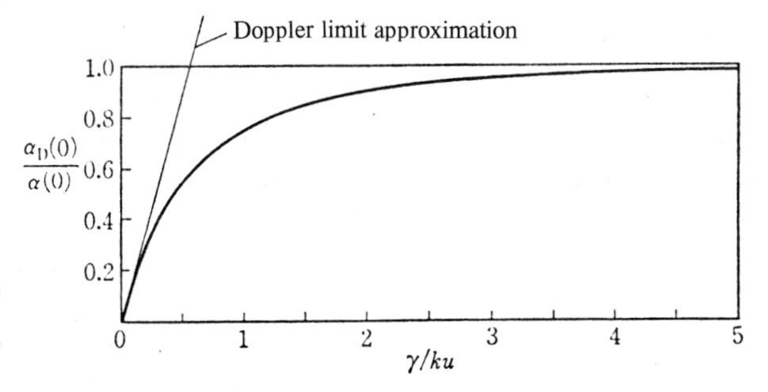

Fig. 8.5. Intensity of a spectral line with both Doppler and homogeneous broadening. Since γ is almost proportional to the gas pressure, this curve represents the variation of the absorption constant with pressure

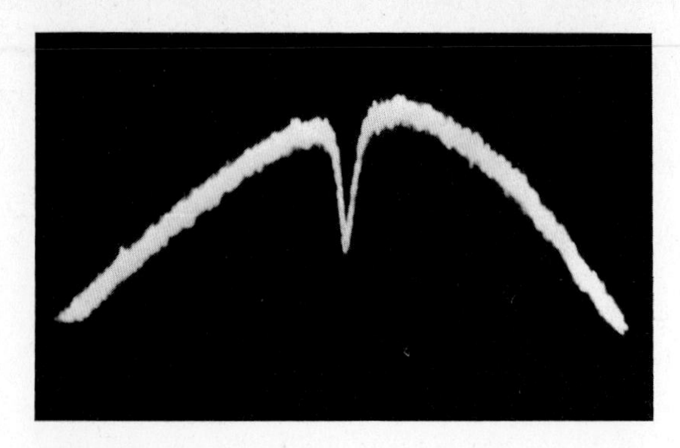

Fig. 8.6. The observed hole-burning effect of a spectral line

over the inhomogeneous width. Thus, when an inhomogeneously broadened spectral line is saturated by absorbing strong incident light at a certain frequency, the absorption of weak probe beam, as its frequency is scanned, is observed to exhibit a dip in the absorption profile, as shown in Fig. 8.6. The effect of burning a hole in a spectral line is called the hole-burning effect. The hole-burning effect in a Doppler broadened spectral line of a gas will be discussed in some detail in the following [8.3].

As stated in the preceding section, the resonance frequency ω_0 is shifted to $\omega_0 + kv$ because of the Doppler effect, where v is the velocity component of the gaseous molecule in the direction of the incident light. Thus, molecules with different velocity components have different transition probabilities. The absorption cross-section or the average transition probability for a molecule with velocity component v is obtained by replacing ω_0 in (8.14 or 8.2) by $\omega_0 + kv$. The absorption cross-section of the moving molecule can now be expressed as a function of ω and v in the form

$$\sigma(v) = \frac{k\gamma|\mu_{12}|^2}{\varepsilon_0\hbar[(\omega_0 + kv - \omega)^2 + \gamma^2]} \tag{8.52}$$

and the transition probability for the incident photon flux I becomes

$$I\sigma(v) = \frac{|x|^2}{2} \cdot \frac{\gamma}{(\omega_0 + kv - \omega)^2 + \gamma^2} . \tag{8.53}$$

Let $N_1(v)$ and $N_2(v)$ represent the molecular velocity distributions in the lower and upper levels of the transition. Then the velocity distributions of two-level atoms in the steady-state saturated-absorption can be expressed, by rewriting (8.21), as

$$\begin{aligned} N_1(v) &= N_1^{(0)}(v) - \frac{\gamma}{2\gamma_1} \cdot \frac{[N_1^{(0)}(v) - N_2^{(0)}(v)]|x|^2}{(\omega_0 + kv - \omega)^2 + (\Delta\omega_h)^2} , \\ N_2(v) &= N_2^{(0)}(v) + \frac{\gamma}{2\gamma_2} \cdot \frac{[N_1^{(0)}(v) - N_2^{(0)}(v)]|x|^2}{(\omega_0 + kv - \omega)^2 + (\Delta\omega_h)^2} , \end{aligned} \tag{8.54}$$

where the meanings of terms not containing v are the same as before. Assuming the Maxwell-Boltzmann molecular velocity distribution in each state when there is no incident light, we have

$$N_1^{(0)}(v) = \frac{N_1^{(0)}}{\sqrt{\pi}\,u}\,\mathrm{e}^{-v^2/u^2}\,, \qquad N_2^{(0)}(v) = \frac{N_2^{(0)}}{\sqrt{\pi}\,u}\,\mathrm{e}^{-v^2/u^2}\,, \tag{8.55}$$

which are shown by dotted lines in Fig. 8.7.

When the incident light is sufficiently strong to saturate the absorption, the number of molecules in the lower level decreases and the number in the upper level increases. The increase and decrease in the number of molecules are not uniform if the line is inhomogeneously broadened. In the case of a Doppler broadened line, the increase and decrease are the largest for molecules whose velocity component is

$$v = \frac{\omega - \omega_0}{k}\,,$$

as indicated by the second term on the right-hand side of (8.54). The increase and decrease are smaller than half their maximum values, when v differs by more than $\Delta\omega_{\mathrm{h}}/k$ from the value above. Therefore, when strong light with frequency ω is incident, the velocity distributions for the two levels are like those shown by full lines in Fig. 8.7. Then the population difference between the lower and upper levels is calculated to be

$$N_1(v) - N_2(v) = \frac{N_1^{(0)}(v) - N_2^{(0)}(v)}{1 + 2\tau I\sigma(v)} \tag{8.56}$$

where τ is the longitudinal relaxation time (8.20).

The incident power absorbed per unit volume by a gas consisting of two-level molecules with such velocity distributions is given by

$$\Delta P = \hbar\omega \int_{-\infty}^{\infty} [N_1(v) - N_2(v)]\, I\sigma(v)\, dv\,. \tag{8.57}$$

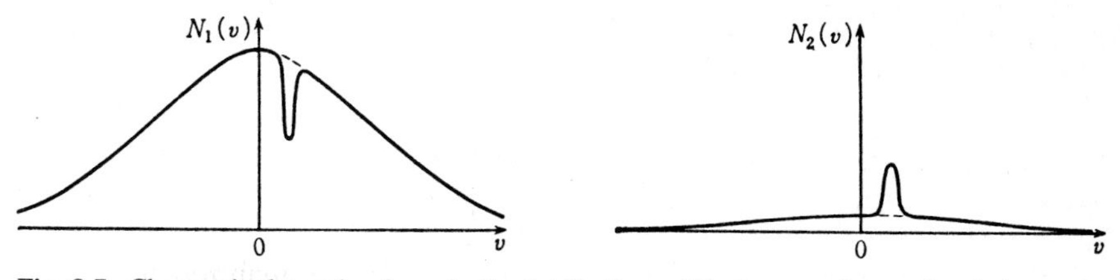

Fig. 8.7. Changes in the molecular velocity distributions of the lower and upper levels due to the hole-burning effect

Therefore, from (8.57, 56), the power absorption coefficient $2\alpha_D$ can be expressed as

$$2\alpha_D = \frac{\Delta P}{\hbar\omega I} = \int_{-\infty}^{\infty} \frac{N_1^{(0)}(v) - N_2^{(0)}(v)}{\sigma(v)^{-1} + 2\tau I} dv\,. \tag{8.58}$$

Substitution of (8.17, 52, 55) into (8.58) gives exactly the same equation as (8.41).

However, if a weak probe light of frequency ω' is introduced in addition to the strong saturating light, a dip in the absorption line will be observed, as shown in Fig. 8.6, as ω' is scanned. This is because the intensity of the absorption is proportional to $N_1(v) - N_2(v)$, in which v is related to the probe frequency ω' by $v = (\omega' - \omega_0)/k$.

A similar dip in emission will appear in the emission line of a medium with an inverted population. The central frequency of the dip observed in this hole-burning experiment is at $\omega' = \omega$. As ω' is scanned, the width of the dip is not γ but 2γ, or, more accurately, $\gamma + \Delta\omega_h$. The reason for this is that, even if the hole burning in the molecular velocity distribution appeared to be as sharp as the Dirac delta function, the observed width probed with the frequency ω' would be γ. Therefore, the observed width in the hole-burning experiment is the sum of the width of the hole burnt by the strong light at ω and the linewidth of the probe absorption at ω'.

8.6 Coherent Transient Phenomena

So far, we have mainly dealt with the steady state of the coherent interaction between the incident light and the two-level medium. Even though it is not strictly a steady-state phenomenon, provided the time variation is sufficiently slow compared with the relaxation time of the medium, the interaction may be well approximated as steady state. Normally, the relaxation time for solids and liquids at room temperatures is of the order of picoseconds or shorter; therefore, the interaction in most cases may be treated as quasi-steady. However, even for solids, there are exceptions where the relaxation time is long, such as in laser transitions, etc., while the relaxation time in gases can be fairly long when the pressure is low. In such media we may observe various transient phenomena of the coherent interaction without necessarily using picosecond pulses. Observations of coherent transient effects are useful in studying the dynamic properties of excited states, such as the relaxation process in a medium, the transfer of excitation energy to some other state, and chemical reactions. Moreover, they are important as foundations for various high-speed devices in opto-electronics.

Most of the coherent transient phenomena can be analyzed to a good approximation by using the density-matrix formalism, in which the relaxation in the medium of two-level molecules is phenomenologically taken into account [8.4]. Even if the approximation were found to be unsatisfactory, it would not be easy to develop a general theory which gives a better approximation.

We have so far discussed the response of a medium when the incident light is steady and monochromatic. However, when the intensity of the monochromatic incident light changes with time, sideband components of different frequencies will be developed, when rapid changes of the amplitude of the incident light occur; moreover, the amplitude and phase of the optical electric field reacting on the molecules at different positions will be different. Thus, the variation of propagation constants for different frequencies (dispersion of the refractive index) must be taken into consideration when investigating transient phenomena of the nonlinear coherent interaction between two-level molecules and the optical field involving spatial and temporal variations of propagation effects. Furthermore, since the relaxation of the induced dipole of the two-level molecule (transverse relaxation) differs from the relaxation of the population (longitudinal relaxation), the spatial and temporal distributions of the molecular dipole and the population are different. It is the polarization of the medium that governs the propagation of light. Here the polarization is given by the product of the population and the induced dipole moment of the two-level molecule.

Therefore, in the treatment of coherent transient effects, it must be remembered that both the optical electric field $\boldsymbol{E}$ of the light propagating through the medium and the polarization $\boldsymbol{P}$ of the medium itself are rapidly changing functions, both in space and in time. The spatial and temporal variations of the optical electric field $\boldsymbol{E}$ in the presence of the induced polarization $\boldsymbol{P}$ must be expressed according to the Maxwell equations, while the polarization $\boldsymbol{P}$ of the two-level medium, induced by the optical electric field, must be expressed by the equations of motion of the density matrix. Then the solutions for $\boldsymbol{E}$ and $\boldsymbol{P}$ must be obtained consistent with these two kinds of equations at all instants of time and everywhere in space. However, it is almost impossible even for a high-speed computer to obtain such perfect solutions in a number of practical problems. Therefore, the method used at present is to obtain approximate solutions by making drastic simplifications and to observe coherent transient phenomena under such experimental conditions as to make theoretical calculations easier.

In an actual medium there exist many transitions between levels other than the two levels in resonance with the incident light, and there may coexist some molecules other than the two-level molecules. We shall denote the polarization which is due only to the transition between the particular two levels by $\boldsymbol{P}$, and the polarization due to other causes by $\varepsilon_0 \chi \boldsymbol{E}$. Here we shall assume that χ is a linear susceptibility, being independent of the light intensity, and that the nonlinear coherent interaction is included in $\boldsymbol{P}$. By putting $\varepsilon_0(1 + \chi) = \varepsilon$, the electric displacement containing both polarizations is expressed as

$$D = \varepsilon E + P \ . \tag{8.59}$$

This is substituted into the wave equation (3.7) derived from the Maxwell equations, to give

$$\nabla \times \nabla \times \boldsymbol{E} + \varepsilon\mu_0 \frac{\partial^2 \boldsymbol{E}}{\partial t^2} = -\mu_0 \frac{\partial^2 \boldsymbol{P}}{\partial t^2} \ . \tag{8.60}$$

Since it is of great mathematical difficulty to analyze nonlinear coherent phenomena for three-dimensional waves, we shall restrict our investigations to one-dimensional waves. Taking the z axis in the direction of the light propagation, we express $\boldsymbol{E}$ and $\boldsymbol{P}$ for plane waves as

$$\begin{aligned} \boldsymbol{E}(z,t) &= \tfrac{1}{2}\,\mathscr{E}(z,t)\exp(\mathrm{i}\omega t - \mathrm{i}kz) + \text{c.c.}\ , \\ \boldsymbol{P}(z,t) &= \tfrac{1}{2}\,\mathscr{P}(z,t)\exp(\mathrm{i}\omega t - \mathrm{i}kz) + \text{c.c.}\ , \end{aligned} \tag{8.61}$$

where ω is the frequency of the light wave, and the wavenumber k is independent of the transition between the two levels. It is defined by

$$k = \omega\sqrt{\varepsilon\mu_0} = \eta\frac{\omega}{c} \ , \tag{8.62}$$

where $\eta = \sqrt{\varepsilon/\varepsilon_0}$ is the refractive index of the medium. Then $\mathscr{E}(z,t)$ and $\mathscr{P}(z,t)$ are the envelopes of $\boldsymbol{E}$ and $\boldsymbol{P}$, respectively, which vary slowly in time compared to ω and gradually in space compared to k.

Using (8.61), the wave equation (8.60) is rewritten as

$$\frac{\partial^2 \boldsymbol{E}(z,t)}{\partial z^2} - \varepsilon\mu_0 \frac{\partial^2 \boldsymbol{E}(z,t)}{\partial t^2} = \mu_0 \frac{\partial^2 \boldsymbol{P}(z,t)}{\partial t^2} \ . \tag{8.63}$$

Abbreviating the notations $\boldsymbol{E}(z,t)$ and $\mathscr{E}(z,t)$ to $\boldsymbol{E}$ and $\mathscr{E}$, we have

$$\frac{\partial^2 \boldsymbol{E}}{\partial z^2} = \frac{1}{2}\left(\frac{\partial^2 \mathscr{E}}{\partial z^2} - 2\mathrm{i}k\frac{\partial \mathscr{E}}{\partial z} - k^2\mathscr{E}\right)\exp(\mathrm{i}\omega t - \mathrm{i}kz) + \text{c.c.}$$

If the spatial variation of $\mathscr{E}$ is so gradual compared to k that

$$\left|\frac{\partial \mathscr{E}}{\partial z}\right| \ll k|\mathscr{E}| \ ,$$

then the second-order derivative $\partial^2 \mathscr{E}/\partial z^2$ can be neglected. Similarly, for the time derivatives we may assume

$$\left|\frac{\partial \mathscr{E}}{\partial t}\right| \ll \omega|\mathscr{E}| \ ,$$

and neglect the term $\partial^2 \mathscr{E}/\partial t^2$, since the temporal variation of $\mathscr{E}$ has been taken to be slow compared to ω. Such an approximation is called the slowly varying envelope approximation, abbreviated to SVEA. The equation obtained by differentiating the polarization $\boldsymbol{P}$ twice with respect to time is similar to the above; but $\boldsymbol{P}$ is a perturbation in the wave equation and is a small quantity compared to $\varepsilon \boldsymbol{E}$ or $\varepsilon_0 \boldsymbol{E}$, so that not only $\partial^2 \mathscr{P}/\partial t^2$, but $\partial \mathscr{P}/\partial t$ also can be neglected. The wave equation (8.63) is then written in the SVEA as

$$\frac{\partial \mathscr{E}}{\partial z} + \frac{\eta}{c}\frac{\partial \mathscr{E}}{\partial t} = -\frac{\mathrm{i}k}{2\varepsilon}\mathscr{P} . \tag{8.64}$$

If $\mathscr{P} = 0$, the right-hand side of (8.64) vanishes and this wave equation indicates that $\mathscr{E}(z, t)$ travels without change in its waveform with a velocity c/η in the $+z$ direction. Thus, if we put

$$\tau = t - \frac{\eta}{c} z , \tag{8.65}$$

the waveform at an arbitrary point z can be expressed by $\mathscr{E}(z, t) = \mathscr{E}(\tau)$.

When $\mathscr{P} = 0$, the waveform $\mathscr{E}(\tau)$ is unchanged during propagation at the velocity c/η. When $\mathscr{P} \neq 0$, on the contrary, the waveform changes as it progresses. The change in velocity can also be expressed, in general, as a change in waveform. In the case when $\mathscr{P}$ is sufficiently small and the length of the medium is not too long, the higher-order interactions need not be taken into account and the analysis is much simplified. Then we consider that there is only a small difference between the input waveform at $z = 0$ and the output waveform at $z = L$. If $\Delta\mathscr{E}$ denotes the increase in $\mathscr{E}(\tau)$ between $z = 0$ and $z = L$, we obtain from (8.64, 65)

$$\frac{\Delta \mathscr{E}}{L} \simeq -\frac{\mathrm{i}k}{2\varepsilon}\mathscr{P} , \tag{8.66 a}$$

therefore,

$$\Delta \mathscr{E} = -\frac{\mathrm{i}kL}{2\varepsilon}\mathscr{P} . \tag{8.66 b}$$

Thus, if we observe the change in the output waveform in such a case, we will find the temporal variation of the nonlinear coherent polarization of the two-level medium. Various coherent transient effects are being investigated by using stepwise- or pulse-modulated laser light so as to simplify the comparison between the experimental and the corresponding theoretical results. Figure 8.8 depicts typical effects such as optical nutation, free induction decay, photon echo, and self-induced transparency, which are explained in the following. A more detailed treatment can be found in [8.5].

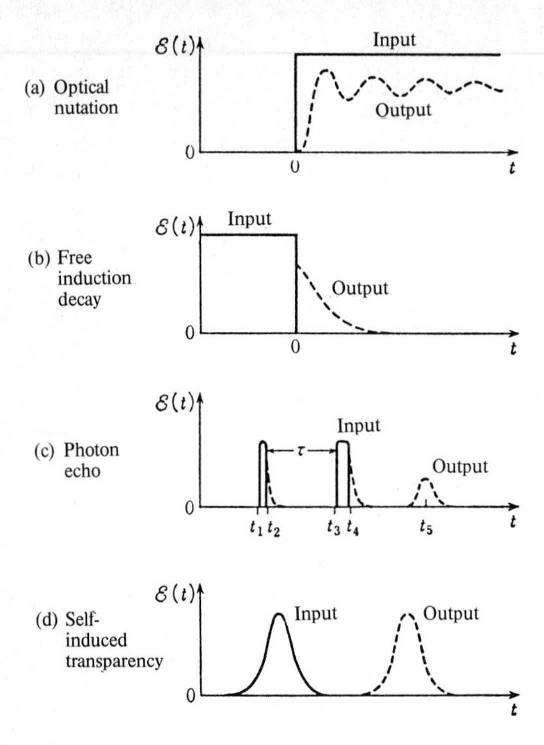

Fig. 8.8 a–d. Schematic diagrams of typical coherent transient phenomena. (**a**) Optical nutation. (**b**) Free induction decay. (**c**) Photon echo. (**d**) Self-induced transparency

8.6.1 Optical Nutation

As stated in Sect. 7.1, when incident light of constant frequency and amplitude is applied at $t = 0$ to two-level molecules that were unperturbed for $t < 0$, the absorption by the molecules exhibit a nutation effect shown in Fig. 7.1. In order to distinguish this effect from nutations in a spinning top and magnetic resonance, it is called optical nutation. Because of relaxation and line-broadening effects, the optical nutation actually observed takes the form of damped oscillations, as shown in the dotted line in Fig. 8.8 a.

In a homogeneous medium, damping of the optical nutation signal is mainly determined by the transverse relaxation time. In a Doppler broadened transition, however, the resonance frequency of each molecule in the medium is slightly different, and the corresponding nutation frequency (7.15) differs, as shown by the two curves in Fig. 7.1. As a result, the optical nutation signal of all the molecules with inhomogeneous broadening, taken collectively, is destroyed very rapidly and the damping time becomes short. Since the nutation signal from molecules having a large frequency shift is relatively small, the damping time is a little longer than the reciprocal of the inhomogeneous width and is also dependent on the intensity of the incident light.

8.6.2 Free Induction Decay

In contrast to the optical nutation experiment, if light of constant intensity is incident up to the time $t = 0$, and either the amplitude of the incident light is turned off or the frequency is shifted by some amount at $t = 0$ as shown in Fig. 8.8b, the dipole moment induced up to $t = 0$ will radiate and decay in the free unperturbed state for $t > 0$. This is called free induction decay, abbreviated to FID [8.6]. The FID signal is coherent, similar to the incident laser beam, and is emitted in the same direction as the incident beam. This is because the spatial distribution of the phases of the induced dipoles is the same as that of the incident beam.

Using the pseudovector representation of the density matrix in a rotating frame of reference, the behavior of a homogeneous medium, in which the induced dipole moments produced while $t < 0$ are left unperturbed for $t > 0$, can be expressed from (7.61) as

$$\begin{aligned} \frac{du}{dt} &= -\gamma u - (\omega_0 - \omega)\, v\,, \\ \frac{dv}{dt} &= (\omega_0 - \omega)\, u - \gamma v\,, \\ \frac{dw}{dt} &= -\frac{w - w^{(0)}}{\tau}\,. \end{aligned} \tag{8.67}$$

Solving this, we obtain

$$u(t) + \mathrm{i}v(t) = [u(0) + \mathrm{i}v(0)]\, \mathrm{e}^{\mathrm{i}(\omega_0 - \omega)t}\, \mathrm{e}^{-\gamma t} \tag{8.68}$$

where the initial values $u(0)$ and $v(0)$ represent the induced dipoles at $t = 0$ and are given by (8.28). Since (8.68) is expressed in a frame rotating with the frequency ω with respect to the laboratory frame, the radiating dipoles of FID are oscillating with the proper frequency ω_0 and the decay time is equal to the transverse relaxation time $1/\gamma$.

Let the frequency of the incident light be suddenly switched to ω', without its amplitude changing, so that the frequency shift is larger than the linewidth, i.e., $|\omega' - \omega_0| \gg \gamma$. Then a beat between the incident light of frequency ω' and the FID signal at ω_0 will appear for the subsequent time $t > 0$. This is called the free induction beat. Since observation of this beat is equivalent to the heterodyne detection of the FID signal, it is easier than the detection of the FID power emitted when the incident light is suddenly turned off. Figure 8.9 shows a FID beat observed for the 10.8 μm vibration-rotation transition of $^{15}NH_3$ at 0.87 Pa, using an N_2O laser.

The decay time of FID in a homogeneous medium, like that of the optical nutation, is equal to the transverse relaxation time. However, as in the case above, when the line is Doppler broadened, the FID frequencies of molecules of different velocities are different. Due to dephasing among these

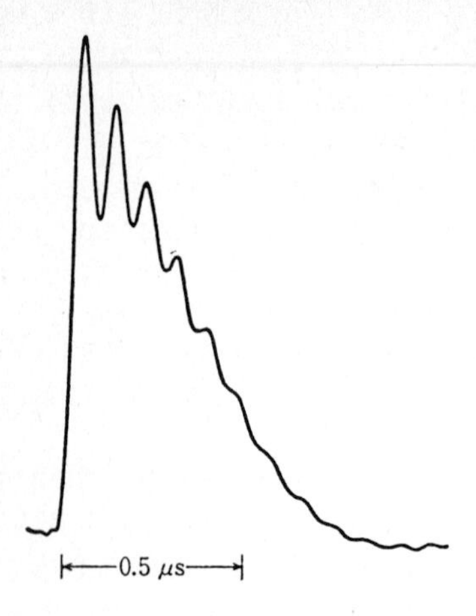

Fig. 8.9. Free induction beat observed in an infrared spectral line of ammonia. The N_2O P(15) laser line of intensity 45 mW cm^{-2} is incident on $^{15}NH_3$ at 0.87 Pa pressure. (By courtesy of Dr. Tadao Shimizu)

molecules, the resultant beat signal dies out more rapidly. The shortening of the decay time in this case is not only due to inhomogeneous broadening, but also depends on the light intensity for the following reason. It is the group of molecules involved in hole burning that dominantly contributes to the FID signal; these molecules are situated within a narrow homogeneous width, somewhere within the whole range of inhomogeneous broadening. As the intensity and the width of hole burning are dependent on the light intensity, so is the FID. Furthermore, the decay time of the observed FID signal also becomes shorter due to the spatial intensity distribution of the laser beam and the transverse motion of molecules.

Now, among the collisions of molecules in a gas, there are weak collisions[6] which do not change the magnitude of the energy and momentum of the molecules but change the phase of the molecular dipole moment or the direction of the molecular velocity. These are called phase-changing collisions and small-angle collisions, respectively. Because of such collisions, the phases of the FID signals emitted by the molecules become disordered and the decay time of the FID due to the gaseous molecules as a whole is shortened. Such effects also influence the free induction beats, in a somewhat different way. Thus, it is possible to study these weak collision processes. It is difficult, however, to separate all these processes, in general.

8.6.3 Photon Echo

Unlike optical nutation and free induction decay, the photon echo appears only with an inhomogeneously broadened line. When two light pulses are incident separated by a time interval τ, a light pulse is emitted at time τ after the second pulse, like a light echo. This phenomenon is usually called the

---

6 Also known as soft collisions.

photon echo. As will be explained in the following, the echo is generated by a mechanism similar to that of the spin echo, that is, an interference effect of light waves; since the effect is not due to the corpuscular nature of light, there is actually no reason why it should be called the photon echo, and it is sometimes referred to as the "photo echo".

Before explaining the generation of the photon echo, it should be noted that the effect of a light pulse on a two-level molecule can be expressed, in general, by an angle of rotation of the pseudovector of the density matrix. The light pulse incident on a two-level molecule is expressed as

$$E(t) = \tfrac{1}{2}\mathscr{E}(t)\,\mathrm{e}^{\mathrm{i}\omega t} + \text{c.c.} \tag{8.69}$$

where $\mathscr{E}(t)$ represents the time-varying envelope of the amplitude. At resonance, when $\omega = \omega_0$, the pseudovector of the density matrix $\boldsymbol{\varrho}$ in the rotating frame, as described in Sect. 7.5, will swing round at the nutation frequency $|\mu_{12}\mathscr{E}(t)|/\hbar$ in a plane containing the Z axis. Then, the angle of nutation from $t = t_1$ to t_2 is given by

$$\Theta = \frac{|\mu_{12}|}{\hbar}\int_{t_1}^{t_2} |\mathscr{E}(t)|\,dt\,. \tag{8.70}$$

Taking the time origin so as to make the phase of the incident light zero, $\mathscr{E}(t)$ is a real quantity, which can be represented by a vector in the direction of the X' axis. Then the pseudovector $\boldsymbol{\varrho}$ rotates in the clockwise direction through an angle Θ in the $Y'Z'$ plane. If $\mathscr{E}(t) = 0$ for $t < t_1$ and $t > t_2$, Θ is the angular difference of the vector $\boldsymbol{\varrho}$ before and after the passage of the pulse. A pulse for which $\Theta = \pi\,(180°)$ is called a π-pulse or a 180-degree pulse, and that for which $\Theta = \pi/2$ (90°) is called a $\pi/2$-pulse or a 90-degree pulse. Since Θ represents the integral of the pulse amplitude $\mathscr{E}(t)$, as given by (8.70), it is sometimes called the pulse area.

Now, although the photon echo occurs, in general, when two pulses of arbitrary intensities are incident, the maximum echo intensity is obtained when the first pulse is a $\pi/2$-pulse and the second is a π-pulse. Formation of the photon echo, when a $\pi/2$-pulse and a π-pulse are incident sequentially, is illustrated in Fig. 8.10 using the vector model in the rotating frame. Here the time sequence $t = t_1, t_2, \ldots$ corresponds to $t = t_1, t_2, \ldots$ in Fig. 8.8c. Even when the line is inhomogeneously broadened to the inhomogeneous width $\Delta\omega_0$, we may assume $\Delta\omega_0 \cdot \Delta t \ll 1$ if the pulse width Δt is short enough. In such a case, the effect of the $\pi/2$-pulse is to rotate the pseudovectors of the molecules with inhomogeneous broadening through 90° about the X' axis.

As shown in Fig. 8.10, the pseudovectors $\boldsymbol{\varrho}$ of all molecules, initially in the $-Z'$ direction at $t = t_1$, are directed toward the Y' axis at $t = t_2$ immediately after the $\pi/2$-pulse. Subsequently, the pseudovector $\boldsymbol{\varrho}$ of each molecule precesses at a different angular velocity from ω, because of the effect of inhomogeneous broadening. This results in dephasing at $t = t_3$, after a lapse of time $\tau \gg (\Delta\omega_\mathrm{D})^{-1}$, as shown by different vectors directed to A, B, C, ... in

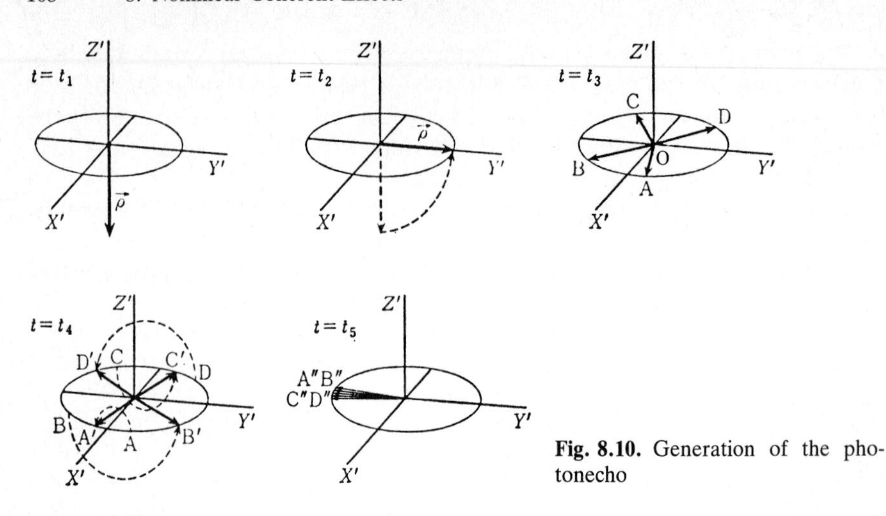

Fig. 8.10. Generation of the photonecho

the $X'Y'$ plane. At this moment the ensemble average of $\boldsymbol{\varrho}$ has practically disappeared. When the π-pulse is applied, the pseudovectors $\boldsymbol{\varrho}$ of the molecules are turned through 180 degrees around the X' axis so that they take directions shown by A′, B′, C′, ... at $t = t_4$, the ensemble average of $\boldsymbol{\varrho}$ still being unseen. As time elapses after t_4, these pseudovectors rephase to line up along the $-Y'$ axis at $t_5 = t_4 + \tau$, as shown by A″, B″, C″, ... The reason for rephasing is that each molecule keeps the same precession frequency before and after the π-pulse, so that the phase angle of each $\boldsymbol{\varrho}$ is given by

$$\angle \mathrm{AOY}' = \angle \mathrm{A}''\mathrm{OA}' , \quad \angle \mathrm{BOY}' = \angle \mathrm{B}''\mathrm{OB}' , \ldots$$

as can be seen in Fig. 8.10.

It is in this manner that, after a lapse of time τ following the second pulse, the ensemble average of $\boldsymbol{\varrho}$ is reproduced to emit a light pulse from the constructive combination of coherent dipoles. This is the photon echo. In an actual medium in which relaxation is inevitable, if the pulse interval τ is made longer than the relaxation time, the photon echo becomes gradually invisible. This decay is mainly determined by the phase relaxation time of the dipole moment, namely, its transverse relaxation time.

8.6.4 Self-Induced Transparency

Among a variety of coherent transient phenomena other than those mentioned above, one of the most remarkable phenomena in connection with propagation is known as self-induced transparency (abbreviated to SIT). When a 2π-pulse is incident on an absorbing two-level medium, the pulse passes through the medium without being absorbed, provided the pulse width is shorter than the relaxation time by a sufficient amount. Since

$\Theta = 2\pi$ when the pulse area is 2π, the vector $\boldsymbol{\varrho}$ representing the density matrix of the two-level medium rotates through 360°. While the vector $\boldsymbol{\varrho}$ turns initially from 0° to 180°, the incident light is absorbed; then, as it turns from 180° to 360°, the same amount of absorbed light is emitted. Therefore, the incident light can pass through the absorbing medium without being essentially absorbed. This is self-induced transparency, which differs instrinsically from saturated absorption.

Self-induced transparency is a phenomenon where a 2π-pulse, which is shorter than the relaxation time of the absorbing medium, is almost 100% transmitted through the medium. In saturated absorption, on the other hand, the absorbance decreases gradually with the incident light intensity whose amplitude is constant or varies slowly compared with the relaxation time of the absorber. Hence the transmittance in saturated absorption cannot be 100%. In the case of self-induced transparency, the two-level molecules in the lower level at different parts of the medium interact coherently with the incident light pulse, and, once having absorbed it, coherently emit the energy thereafter. Therefore, the velocity of propagation of the 2π-pulse becomes less than one-tenth or even one-hundredth of that of ordinary light.

Such a phenomenon can be analyzed using the approximate equation (8.64) derived from the Maxwell equations for the propagation of plane waves, where the polarization of every part of the medium is given by the optical Bloch equations (7.61). When the light pulse is shorter than the relaxation time, the calculation is simplified, because we may then neglect the relaxation term by putting $\gamma = 0$ and $\tau = \infty$ in (7.61). However, one must take into consideration the fact that the frequency components of a narrow pulse are widely spread and that molecules in different places behave differently in time.

Such theoretical calculations have revealed the following. When a 2π-pulse is incident on an absorbing medium consisting of two-level molecules, the pulse area remains unchanged, though the transmitted waveform may change somewhat. Moreover, when the pulse area of the incident light lies between π and 3π, the pulse area of the transmitted light approaches 2π. Besides the 2π-pulse, pulses of area 4π, 6π, ... may also propagate in a similarly stable manner. In particular, when the envelope $\mathscr{E}(t)$ of the amplitude of a 2π-pulse takes the form of a hyperbolic secant function as shown in Fig. 8.11, the 2π-pulse will travel with a constant speed without changing its waveform and amplitude.

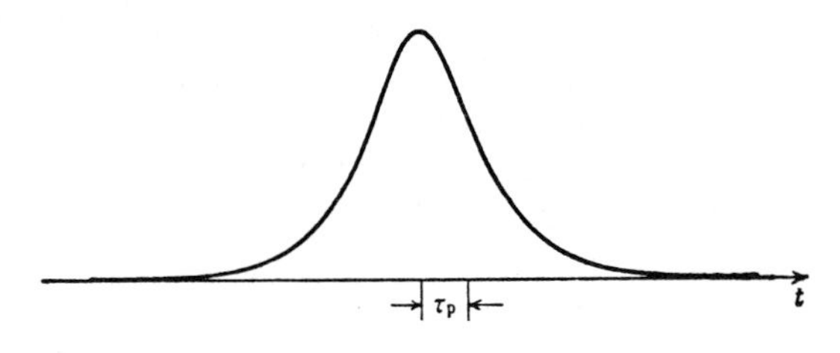

Fig. 8.11. Profile of the hyperbolic secant function sech (t/τ_p)

The amplitude $\mathscr{E}(z,t)$ of such a 2π-pulse traveling in the z direction is expressed as [8.5]

$$\frac{|\mu_{12}|}{\hbar}\mathscr{E}(z,t) = \frac{2}{\tau_p}\operatorname{sech}\left(\frac{t}{\tau_p} - \frac{z}{v_p\tau_p}\right), \tag{8.71}$$

where v_p is the velocity of propagation and τ_p is the pulse width, i.e., the halfwidth where the amplitude becomes $2/(e + e^{-1}) = 0.648$ times the maximum. The velocity v_p of the pulse traveling with self-induced transparency is given by [8.5]

$$\frac{1}{v_p} = \frac{\eta}{c}\left(1 + \frac{N\omega|\mu_{12}|^2\tau_p^2}{2\varepsilon\hbar}\int_{-\infty}^{\infty}\frac{g(\omega')\,d\omega'}{1+(\omega-\omega')^2\tau_p^2}\right). \tag{8.72}$$

Here $\eta = \sqrt{\varepsilon/\varepsilon_0}$ is the refractive index of the medium, N (or $N_1^{(0)} - N_2^{(0)}$, if the upper level is populated) is the number of molecules per unit volume, and $g(\omega)$ is the normalized lineshape function of the transition between the two levels. If the absorption linewidth $\Delta\omega_0$ is so narrow that $\Delta\omega_0 \cdot \tau_p \ll 1$, this integral is equal to 1. On the contrary, if $\Delta\omega_0 \cdot \tau_p \gg 1$, it is equal to $(\pi/\tau_p)\,g(\omega)$.

In the case when the spectral width $\Delta\omega_0$ of the inhomogeneously broadened line is narrow, so that $\Delta\omega_0 \cdot \tau_p \ll 1$, (8.72) may be written in the form

$$\frac{1}{v_p} = \frac{\eta}{c}\left(1 + \frac{U_m}{U_{em}}\right), \tag{8.73}$$

where U_{em} is the peak energy of the light pulse per unit volume and $U_m = N\hbar\omega$ is the maximum energy of the two-level molecules. Because the peak value of the optical electric field is given by

$$\mathscr{E}_m = \frac{2\hbar}{|\mu_{12}|\,\tau_p},$$

the energy density of the light pulse becomes

$$U_{em} = \frac{1}{2}\varepsilon E^2 \times 2 = \frac{1}{2}\varepsilon\mathscr{E}_m^2 = \frac{2\varepsilon\hbar^2}{|\mu_{12}|^2\tau_p^2}.$$

Equation (8.73) implies that the reason for the slowing down of the propagation velocity in self-induced transparency is that the optical pulse energy is temporarily transferred to the excitation energy of the two-level molecules.

Problems

8.1 Rewrite the amplitude absorption coefficient (8.7) as a function of the frequency ω and the power density P of the incident light in terms of the linear absorption constant at the line center $\alpha(0)$ and the saturation power P_s.

Answer: $\alpha(\omega, P) = \gamma^2 \alpha(0)\,[(\omega_0 - \omega)^2 + \gamma^2(1 + P/P_s)]^{-1}$.

8.2 Calculate the absorption cross-section of an atom for its resonance radiation at a wavelength of 1 μm if it has a transition dipole moment of 1 debye and a fluorescence lifetime of 10^{-9} s. Doppler broadening may be neglected.

Answer: $7.5 \times 10^{-17}\,\mathrm{m}^2 = 7.5 \times 10^{-13}\,\mathrm{cm}^2$.

8.3 Explain the physical reason why the small-signal absorption constant is independent of the longitudinal relaxation time.

Hint: Mathematically, it can be referred to (8.23).

8.4 Find a formula for the saturation value of the absorbed power ΔP_s in terms of the unperturbed population difference when the longitudinal and transverse relaxation constants are different.

Answer: $\Delta P_s = (N_1^{(0)} - N_2^{(0)})\,\hbar\omega/2\tau$.

8.5 Show that the locus of the nonlinear susceptibility (8.31), when the parameter ω changes, is an ellipse in the complex plane. Then find the direction of the major axis and the eccentricity of the ellipse.

Answer: Its direction is along the real axis (χ'), and the eccentricity is $e = \sqrt{\tau|x|^2/(\gamma + \tau|x|^2)}$.

8.6 Find the wavelength at which the natural width equals the Doppler width at 300 K for an atom having a transition dipole moment of 1 debye and an atomic weight of 100.

Hint: Use (4.31) and (8.39).

Answer: Soft x-ray at $\lambda = 1.64 \times 10^{-8}$ m = 16.4 nm.

8.7 Derive (8.51) and show that P_s in (8.51) is identical to that given by (8.24)

8.8 Consider a hole-burning experiment as described in Sect. 8.5. What will be the width of the hole observed by a probe beam making a small angle θ with the saturating beam.

Answer: The observed profile of the hole will be a convolution of Lorentzian broadening of halfwidth γ and Gaussian broadening of halfwidth $\sqrt{\ln 2}\,\theta ku$. The resultant width for $\theta \ll 1$ is approximately $\gamma + 0.4\,\theta ku$.

8.9 Show from (8.57) that the power absorbed by an inhomogeneously broadened line does not show such an evident saturation as in the case of absorption in the absence of inhomogeneous broadening. Explain the reason for this in terms of hole burning.

Answer: While the depth of the hole saturates as the incident power increases, its width increases more or less linearly with the incident power. Since the absorbed power is proportional to the area of the hole burned in the molecular velocity distribution, it does not saturate until the hole spreads over the whole inhomogeneously broadened line so that $\gamma\tau|x|^2 \cong k^2u^2$.

8.10 Calculate the free induction decay of an ensemble of two-level atoms in the presence of Doppler broadening.

Answer: The average dipole moment decays as $p(t) = p(0)\exp(-\gamma t - k^2u^2t^2/4)\exp(\mathrm{i}\omega_0 t)$ + c.c.

8.11 Explain the reason why the damping time of optical nutation depends on the intensity of the incident light, whereas that of free induction decay is independent of it.

Answer: The saturation effect is involved in optical nutation, but not in FID, which occurs when the incident light is cut off.

8.12 Solve the two-level problem as expressed by (7.12) for a pulse of incident light so that the light amplitude is a function of time. Assume that the incident light is at the resonance frequency $\omega = \omega_0$, and use

$$\theta(T) = \int_{-\infty}^{T} x(t)\,dt = \frac{\mu_{12}}{\hbar}\int_{-\infty}^{T} \mathscr{E}(t)\,dt\,.$$

Answer: $a_1(T) = A\,\mathrm{e}^{\mathrm{i}\theta(T)/2} + B\,\mathrm{e}^{-\mathrm{i}\theta(T)/2}$,

$a_2(T) = A\,\mathrm{e}^{\mathrm{i}\theta(T)/2} - B\,\mathrm{e}^{-\mathrm{i}\theta(T)/2}$,

or linear combinations of $\sin[\theta(T)/2]$ and $\cos[\theta(T)/2]$.

8.13 Calculate the halfwidth at half maximum of the hyperbolic secant pulse given by (8.71).

Answer: HWHM $= \ln(2+\sqrt{3})\,\tau_\mathrm{p} = 1.317\,\tau_\mathrm{p}$.

9. Theory of Laser Oscillation

In Chap. 6 we investigated the output characteristics of the laser with the use of rate equations. In this chapter we shall investigate laser oscillation by taking the phase into consideration, so that the interaction between atoms and laser radiation may be treated not merely as a probability of induced emission but as a coherent interaction. Such theories can be classified into two types: the semiclassical theory in which light is treated as classical electromagnetic waves described by the Maxwell equations, while the behavior of atoms in the electromagnetic field is described quantum-mechanically, and the fully quantum-mechanical theory, in which not only the atoms are quantized but also the light, and both are treated quantum-mechanically.

A physical model of a typical laser is given in Fig. 9.1. As can be seen in this figure, the laser is not a closed system consisting of two-level atoms and an optical resonator, but it is an open system interacting with the external world. It is not in thermal equilibrium, being in contact with hot and cold baths. Moreover, it is characterized by the nonlinear interaction between the light and the atoms, in which both atomic and optical coherences must be considered.

9.1 Fundamental Equations of the Semiclassical Theory

The semiclassical theory of the laser is essentially the same as the theory of the molecular beam maser developed by the author and others in 1955 [9.1].

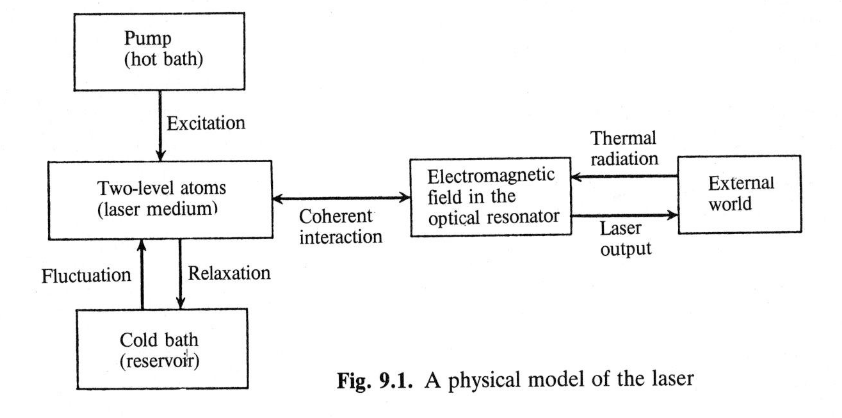

Fig. 9.1. A physical model of the laser

In the semiclassical theory, we first assume a form for the electromagnetic field in each mode of the resonator and quantum-mechanically calculate the polarization of the laser medium induced by the field. Then the Maxwell equations are used to obtain the amplitude and frequency of the electromagnetic waves consistent with both the calculated polarization and the presumed electromagnetic field. In this case, the process of pumping to invert the population of the laser medium and the relaxation that destroys the population inversion are treated as predetermined phenomenological coefficients.

Although there are, in general, an infinite number of modes in an optical resonator, laser oscillation occurs in only a few modes. Even though a single-mode laser oscillating in one mode is desirable, lasers often oscillate in several modes, i.e., they are multimode lasers.

Denoting the mode function of the nth resonance by $U_n(\boldsymbol{r})$ and its proper frequency by Ω_n, we shall distinguish the frequency of the laser oscillation ω_n from Ω_n, because they are not necessarily equal, though they are very close. The optical field $E(\boldsymbol{r}, t)$ and the polarization $P(\boldsymbol{r}, t)$ of a laser medium consisting of two-level atoms can be expanded as

$$E(\boldsymbol{r}, t) = \sum_n \tilde{E}_n(t)\, U_n(\boldsymbol{r})\, \mathrm{e}^{\mathrm{i}\omega_n t} + \mathrm{c.c.}\,, \tag{9.1}$$

$$P(\boldsymbol{r}, t) = \sum_n \tilde{P}_n(t)\, U_n(\boldsymbol{r})\, \mathrm{e}^{\mathrm{i}\omega_n t} + \mathrm{c.c.} \tag{9.2}$$

Although the optical field and the polarization are both vector quantities, we shall restrict our treatment, for the sake of simplicity, to the case of isotropic media so that we do not have to use vector notation. Thus, $\tilde{E}_n(t)$ and $\tilde{P}_n(t)$ denote the magnitudes of the optical field and the polarization, respectively, both of which change much more slowly in time than the optical frequency.

Now, if κ_n denotes the damping constant of the amplitude of oscillation of the nth mode, the free oscillation of the optical field is written as

$$E_n(\boldsymbol{r}, t) = \tilde{E}_n U_n(\boldsymbol{r}) \exp(\mathrm{i}\Omega_n t - \kappa_n t) + \mathrm{c.c.}\,, \tag{9.3}$$

where $\tilde{E}_n$ is a constant. The wave equation (3.7), derived from the Maxwell equations for $\mu = \mu_0$ and $\operatorname{div} \boldsymbol{E} = 0$, becomes

$$\nabla^2 \boldsymbol{E} = \mu_0 \frac{\partial^2 \boldsymbol{D}}{\partial t^2} \tag{9.4}$$

so that, by substituting $\boldsymbol{D} = \varepsilon_0 \boldsymbol{E}$ and (9.3), it can be written as

$$\nabla^2 U_n(\boldsymbol{r}) + \varepsilon_0 \mu_0 \Omega_n^2 U_n(\boldsymbol{r}) + 2\mathrm{i}\varepsilon_0 \mu_0 \kappa_n \Omega_n U_n(\boldsymbol{r}) = 0 \tag{9.5}$$

where the term κ_n^2 has been neglected, since $\kappa_n \ll \Omega_n$.

In the presence of laser, rather than free, oscillation, due to the population inversion of two-level atoms we have $\tilde{P}_n \neq 0$. We therefore substitute

$$\boldsymbol{D} = \varepsilon_0 \boldsymbol{E} + \boldsymbol{P}$$

into (9.4). Substituting (9.1, 2) for $\boldsymbol{E}$ and $\boldsymbol{P}$, and using the orthogonality relation

$$\int U_n^* (\boldsymbol{r}) \, U_m (\boldsymbol{r}) \, d\boldsymbol{r} = \delta_{nm} V_0 \tag{9.6}$$

between the modes[1], we obtain the equation of motion in the form

$$\tilde{E}_n (t) \, \nabla^2 U_n (\boldsymbol{r}) + \varepsilon_0 \mu_0 \left(\omega_n^2 - 2 \mathrm{i} \omega_n \frac{d}{dt} - \frac{d^2}{dt^2} \right) \tilde{E}_n (t) \, U_n (\boldsymbol{r})$$

$$= - \mu_0 \left(\omega_n^2 - 2 \mathrm{i} \omega_n \frac{d}{dt} - \frac{d^2}{dt^2} \right) \tilde{P}_n (t) \, U_n (\boldsymbol{r}) \, .$$

Here, by using the rotating-wave approximation, we have neglected the term $\tilde{E}_n^*$ which changes in time with the factor $\exp(-\mathrm{i}\omega_n t)$. Provided the time variation of $\tilde{E}_n$ and $\tilde{P}_n$ are sufficiently slow, we may also neglect $d^2 \tilde{E}_n / dt^2$, $d^2 \tilde{P}_n / dt^2$, and $d\tilde{P}_n / dt$, as in the approximation used in Sect. 8.6 (SVEA). Thus, the above equation, after substitution of the characteristic equation of the mode (9.5), becomes

$$(\omega_n^2 - \Omega_n^2) \, \tilde{E}_n - 2 \mathrm{i} \kappa_n \Omega_n \tilde{E}_n - 2 \mathrm{i} \omega_n \frac{d\tilde{E}_n}{dt} = - \frac{\omega_n^2}{\varepsilon_0} \tilde{P}_n \, , \tag{9.7}$$

where $\tilde{E}_n (t)$ and $\tilde{P}_n (t)$ are abbreviated here and in the following to $\tilde{E}_n$ and $\tilde{P}_n$. Since $|\omega_n - \Omega_n| \ll \omega_n$, (9.7) can be rewritten as

$$\frac{d\tilde{E}_n}{dt} + [\kappa_n + \mathrm{i} (\omega_n - \Omega_n)] \, \tilde{E}_n = - \frac{\mathrm{i} \omega_n}{2 \varepsilon_0} \tilde{P}_n \, . \tag{9.8}$$

This is an approximation derived from the Maxwell equations representing the relation between the optical field and the polarization of the nth mode.

On the other hand, the polarization and the population inversion of the laser medium, under the perturbation of the optical field given by (9.1), are expressed as $N \varrho_{12} \mu_{21}$ + c.c. and $\Delta N = N \Delta \varrho$, respectively, using the density matrix ϱ described in Chap. 7. Taking into consideration their inhomogeneous distributions in the resonator and variations in time, we shall respectively express them as $P(\boldsymbol{r}, t)$ and $\Delta N (\boldsymbol{r}, t)$. The matrix elements of the perturbation Hamiltonian of the two-level atoms in the optical field (9.1) are given by

$$\mathscr{H}_{12}' = - \mu_{12} \sum_n (\tilde{E}_n \, U_n (\boldsymbol{r}) \, \mathrm{e}^{\mathrm{i} \omega_n t} + \mathrm{c.c.}) \, . \tag{9.9}$$

---

1 δ_{nm} is called the Kronecker delta. Its value is $\delta_{nm} = 0$ when $n \neq m$, and $\delta_{nn} = 1$ when $n = m$. V_0 is a constant with the dimensions of volume.

Then the time variation of the polarization (9.2) can be obtained from (7.49), using (9.9), as

$$\left(\frac{d}{dt} + \gamma - \mathrm{i}\omega_0\right) \sum_n \tilde{P}_n U_n(\boldsymbol{r})\, \mathrm{e}^{\mathrm{i}\omega_n t}$$

$$= \frac{\mathrm{i}}{\hbar} \Delta N(\boldsymbol{r}) |\mu_{12}|^2 \sum_m \tilde{E}_m U_m(\boldsymbol{r})\, \mathrm{e}^{\mathrm{i}\omega_m t} ,$$

where the rotating-wave approximation has been used to separate positive and negative frequency components. Multiplying both sides of this equation by $U_n^*(\boldsymbol{r}) \exp(-\mathrm{i}\omega_n t)$ and integrating over the entire space, we obtain

$$\left(\frac{d}{dt} + \mathrm{i}\omega_n + \gamma - \mathrm{i}\omega_0\right) \tilde{P}_n$$

$$= \frac{\mathrm{i}}{\hbar} |\mu_{12}|^2 \sum_m \tilde{E}_m \int U_n^*(\boldsymbol{r}) \Delta N(\boldsymbol{r})\, U_m(\boldsymbol{r})\, d\boldsymbol{r} \exp[\mathrm{i}(\omega_m - \omega_n)t] .$$

We now put

$$\Delta N_{nm} = \frac{1}{V_0} \int U_n^*(\boldsymbol{r}) \Delta N(\boldsymbol{r})\, U_m(\boldsymbol{r})\, d\boldsymbol{r} \tag{9.10}$$

for the mode expansion coefficient of $\Delta N(\boldsymbol{r})$. Here V_0 is the effective volume of the optical resonator. Though its value varies with the choice of the magnitude of $U_n(\boldsymbol{r})$, it does not affect the results of the calculation. The differential equation for $\tilde{P}_n$ is now rewritten, using (9.10), as

$$\frac{d\tilde{P}_n}{dt} + [\gamma + \mathrm{i}(\omega_n - \omega_0)]\, \tilde{P}_n$$

$$= \frac{\mathrm{i}}{\hbar} |\mu_{12}|^2 \sum_m \Delta N_{nm} \tilde{E}_m \exp[\mathrm{i}(\omega_m - \omega_n)t] . \tag{9.11}$$

Finally, the equation representing the time variation of ΔN_{nm}, the expansion coefficient of the population inversion, can be obtained from (7.48), which expresses the time variation of $\Delta\varrho$. Substituting (9.2, 9) into (7.48) and writing j, k for n, m, we have

$$\frac{d}{dt} \Delta N = -\frac{\Delta N - \Delta N^{(0)}}{\tau}$$

$$- \frac{2\mathrm{i}}{\hbar} \sum_j \sum_k \{\tilde{P}_j^* U_j^*(\boldsymbol{r})\, U_k(\boldsymbol{r})\, \tilde{E}_k \exp[\mathrm{i}(\omega_k - \omega_j)t] - \mathrm{c.c.}\} .$$

Multiplying both sides of this equation by $U_n^*(\boldsymbol{r})\, U_m(\boldsymbol{r})$ and integrating over the space, we obtain

$$\frac{d}{dt}\Delta N_{nm} + \frac{1}{\tau}(\Delta N_{nm} - \Delta N_{nm}^{(0)})$$

$$= \frac{2}{i\hbar}\sum_j \sum_k A_{nm,\,jk}\{\tilde{P}_j^* \tilde{E}_k \exp[i(\omega_k - \omega_j)t] - \text{c.c.}\}\ , \tag{9.12}$$

where we have set

$$A_{nm,\,jk} = \frac{1}{V_0}\int U_n^*(\boldsymbol{r})\,U_m(\boldsymbol{r})\,U_j^*(\boldsymbol{r})\,U_k(\boldsymbol{r})\,d\boldsymbol{r}\ . \tag{9.13}$$

The three equations (9.8, 11, 12) are the fundamental equations of the semiclassical theory of the multimode laser. In deriving (9.8), we have neglected the second derivatives, assuming that the time variation of $\tilde{E}_n$ is not rapid. This assumption is equivalent to the use of the rotating-wave approximation in deriving (9.11). Moreover, a similar approximation has been adopted in (9.12), where the term fluctuating with the difference frequency $\omega_m - \omega_n$ is retained and that fluctuating with the sum frequency $\omega_m + \omega_n$ is neglected, on the assumption that the time variation of ΔN is not rapid.

9.2 Single-Mode Oscillation

Single-mode oscillation can be obtained with a small laser or a laser with a mode-selective resonator. The fundamental equations of a single-mode laser can be written, from (9.8, 11, 12), as

$$\frac{d\tilde{E}}{dt} + [\kappa + i(\omega - \Omega)]\tilde{E} = -\frac{i\omega}{2\varepsilon_0}\tilde{P}\ , \tag{9.14a}$$

$$\frac{d\tilde{P}}{dt} + [\gamma + i(\omega - \omega_0)]\tilde{P} = \frac{i}{\hbar}|\mu_{12}|^2 \Delta N\tilde{E}\ , \tag{9.14b}$$

$$\frac{d}{dt}\Delta N + \frac{1}{\tau}(\Delta N - \Delta N^{(0)}) = \frac{2A}{i\hbar}(\tilde{P}^*\tilde{E} - \tilde{P}\tilde{E}^*)\ . \tag{9.14c}$$

Here the mode subscript has been omitted, since there is only one, and

$$A = \frac{1}{V_0}\int [U^*(\boldsymbol{r})\,U(\boldsymbol{r})]^2\,d\boldsymbol{r}\ . \tag{9.15}$$

When the maximum value of $U(\boldsymbol{r})$ normalized by (9.6) is denoted as $U_{\max}$, the mode volume of the resonator may be defined by

$$V_{\text{mode}} = \frac{V_0}{U_{\max}^2}\ .$$

It is almost equivalent to the volume of the region where the optical field is stronger than average. Although the magnitude of the volume V_0 introduced into (9.6, 10, 15) is arbitrary, it is convenient to assume $V_0 = V_{\text{mode}}$ so that $U_{\max} = 1$.

In the case of multimode oscillation, the polarization $\tilde{P}_n$ of the nth mode is not determined by the electric field $\tilde{E}_n$ of that mode alone, as can be seen from (9.11). In the case of single-mode oscillation, however, the polarization is uniquely determined by the electric field of the mode. It is seen from (9.14 b, c) that $\tilde{P}$ and ΔN must be odd and even functions of $\tilde{E}$, respectively. Therefore, if we write

$$\tilde{P} = \varepsilon_0 \chi \tilde{E} , \tag{9.16}$$

the nonlinear susceptibility χ is an even function of $\tilde{E}$. We write the real part of χ as χ', the imaginary part as $-\chi''$, and $\tilde{E} = |\tilde{E}| \exp(\mathrm{i}\phi)$. Then the equations for the amplitude and phase of a single-mode laser are obtained from the real and imaginary parts of the fundamental equation (9.14 a). These are

$$\left(\frac{d}{dt} + \kappa + \frac{\omega}{2}\chi''\right)|\tilde{E}| = 0 , \quad \text{and} \tag{9.17a}$$

$$\frac{d\phi}{dt} + \omega - \Omega + \frac{\omega}{2}\chi' = 0 , \tag{9.17b}$$

respectively. Since the imaginary part of χ has been conventionally written as $-\chi''$ in order to make χ'' positive for an absorber, χ'' is negative for a medium with population inversion.

9.2.1 Steady-State Oscillation

Let us first obtain the frequency and amplitude of the steady-state oscillation of a single-mode laser. Since $d\tilde{E}/dt = 0$, and $d\tilde{P}/dt = 0$ in the steady state, multiplying the respective sides of (9.14 a and b), and eliminating the common factor $\tilde{P}\tilde{E}$, we obtain

$$[\kappa + \mathrm{i}(\omega - \Omega)][\gamma + \mathrm{i}(\omega - \omega_0)] = \frac{\omega}{2\varepsilon_0 \hbar}|\mu_{12}|^2 \Delta N_{\text{th}} . \tag{9.18}$$

Here, we have written ΔN_{th} for ΔN, because the population inversion in the steady state is equal to the threshold value, as shown in Sect. 6.2. Since the right-hand side of the above equation is real, putting the imaginary part equal to zero gives

$$(\omega - \Omega)\gamma + (\omega - \omega_0)\kappa = 0 , \tag{9.19}$$

and the frequency of oscillation is obtained as

$$\omega = \frac{\Omega\gamma + \omega_0\kappa}{\gamma + \kappa} . \tag{9.20}$$

This equation is the same as (5.53), which was obtained in Chap. 5 for a plane-wave laser with two plane mirrors. Here it has been found to hold not only for plane waves, but also for any resonance mode of variant wave surfaces. This is because the resonance profiles of both the spectral line and the optical resonator are Lorentzian. When the spectral line profile is modified by inhomogeneous broadening, the frequency of laser oscillation will shift with the intensity of oscillation. Needless to say, the laser frequency is subject to change, in accordance with (9.20), with changes in the values of γ and κ, such as the increase of the linewidth γ due to a temperature rise of the medium under strong excitation or a change in the cavity loss κ due to nonlinear optical absorption.

The intensity of laser oscillation is determined by (9.14 c) and the real part of (9.18) as follows. The real part of (9.18) is written

$$\kappa\gamma - (\omega - \Omega)(\omega - \omega_0) = \frac{\omega}{2\varepsilon_0\hbar}|\mu_{12}|^2 \Delta N_{\text{th}} \, .$$

Using (9.19) to eliminate $(\omega - \Omega)$, we obtain the population inversion for steady-state oscillation, that is, the threshold value of the population inversion:

$$\Delta N_{\text{th}} = \frac{2\varepsilon_0\hbar\kappa}{\omega|\mu_{12}|^2\gamma}[(\omega-\omega_0)^2 + \gamma^2] \, . \tag{9.21}$$

This is identical with (6.6), which was obtained from the rate equations, if (4.39) is rewritten from (4.36) as

$$B(\omega) = Bg(\omega) = \frac{|\mu_{12}|^2}{\varepsilon_0\hbar^2} \cdot \frac{\gamma}{(\omega - \omega_0)^2 + \gamma^2} \tag{9.22}$$

when the profile of the spectral line is Lorentzian (4.43).

Since (9.14 a) for the steady state is

$$\tilde{P} = -\frac{2\varepsilon_0}{\mathrm{i}\omega}[\kappa + \mathrm{i}(\omega - \Omega)]\tilde{E} \, ,$$

the right-hand side of (9.14 c) is calculated to be

$$\tilde{P}^*\tilde{E} - \tilde{P}\tilde{E}^* = \frac{4\varepsilon_0\kappa}{\mathrm{i}\omega}|\tilde{E}|^2 \, .$$

Then (9.14 c) gives

$$\Delta N^{(0)} - \Delta N_{\text{th}} = \frac{8\varepsilon_0 A\kappa\tau}{\hbar\omega}|\tilde{E}|^2 \, ,$$

so that we may obtain

$$|\tilde{E}|^2 = \frac{\hbar\omega}{8\,\varepsilon_0 A\kappa\tau}\,(\Delta N^{(0)} - \Delta N_{\mathrm{th}}) \quad \text{or} \tag{9.23}$$

$$|\tilde{E}|^2 = \frac{1}{4\,\varepsilon_0 AB(\omega)\,\tau}\left(\frac{\Delta N^{(0)}}{\Delta N_{\mathrm{th}}} - 1\right) . \tag{9.24}$$

If the optical field intensity within the mode volume V_0 is constant, we have $A = 1$, since $|U(\boldsymbol{r})| = 1$ within this volume. Then we may calculate the energy density W of the laser radiation from (9.1). The result, for the case of a single-mode laser is

$$\begin{aligned} W &= \tfrac{1}{2}\,\varepsilon_0\,|E(\boldsymbol{r},t)|^2 + \tfrac{1}{2}\,\mu_0\,|H(\boldsymbol{r},t)|^2 \\ &= 2\,\varepsilon_0\,|\tilde{E}|^2 . \end{aligned} \tag{9.25}$$

Therefore, when $A = 1$ and $|U(\boldsymbol{r})|^2 = 1$, (9.25) is equivalent to (6.11 a) obtained from the rate equations.

9.2.2 van der Pol Equation

In 1934, *B. van der Pol* [9.2] set up an equation of nonlinear oscillation in order to study the theoretical characteristics of a vacuum tube oscillator. It is known as the van der Pol equation, and is expressed as

$$\frac{d^2x}{dt^2} - (a - bx^2)\,\frac{dx}{dt} + \Omega^2 x = 0 . \tag{9.26}$$

Here x is the amplitude, Ω is the proper frequency of oscillation, and the parameters a and b are ordinarily positive. Not only does this equation describe oscillation in an electronic circuit, it may also describe self-excited oscillations in mechanical, ecological, and other systems. Moreover, by adding an external perturbation term on the right-hand side, it may be extended to describe forced oscillations. In the theory of lasers, the first-order differential equation for the complex amplitude, obtained by using the rotating-wave approximation and SVEA (Sect. 8.6), is used as described in the following.

We shall study the time variation of laser oscillation which is not in a steady state, using the fundamental equations (9.14). Here we assume for brevity that the laser is oscillating at the central frequency of the spectral line. Setting $\Omega = \omega = \omega_0$ and eliminating $\tilde{P}$ by substituting (9.14 a) and its derivative with respect to t into (9.14 b), we obtain

$$\frac{d^2\tilde{E}}{dt^2} + (\kappa + \gamma)\,\frac{d\tilde{E}}{dt} + \kappa\gamma\tilde{E} = \frac{\omega}{2\,\varepsilon_0\hbar}\,|\mu_{12}|^2\,\Delta N\tilde{E} . \tag{9.27 a}$$

Then, $\tilde{P}$ on the right-hand side of (9.14c) can be eliminated by the use of (9.14a) to obtain

$$\frac{d\Delta N}{dt} + \frac{1}{\tau}(\Delta N - \Delta N^{(0)}) = -\frac{4\varepsilon_0 A}{\hbar\omega}\left(\frac{d}{dt}|\tilde{E}|^2 + 2\kappa|\tilde{E}|^2\right). \tag{9.27b}$$

Now, provided

$$\left|\frac{d\tilde{E}}{dt}\right| \ll \gamma|\tilde{E}|, \qquad \left|\frac{d\tilde{E}}{dt}\right| \ll \kappa|\tilde{E}| \tag{9.28}$$

as in the SVEA in Sect. 8.6, we may neglect $d^2\tilde{E}/dt^2$ in (9.27a), and $d|\tilde{E}|^2/dt$ in (9.27b), so that (9.27a, b) become, respectively,

$$(\kappa + \gamma)\frac{d\tilde{E}}{dt} + \kappa\gamma\tilde{E} = \frac{\omega}{2\varepsilon_0\hbar}|\mu_{12}|^2\Delta N\tilde{E} \quad \text{and}$$

$$\tau\frac{d\Delta N}{dt} + \Delta N = \Delta N^{(0)} - \frac{8\varepsilon_0 A\kappa\tau}{\hbar\omega}|\tilde{E}|^2.$$

Eliminating ΔN from these equations, we have

$$\left(\tau\frac{d}{dt} + 1\right)\left[\frac{(\kappa+\gamma)}{\tilde{E}}\frac{d\tilde{E}}{dt} + \kappa\gamma\right] = \frac{\omega}{2\varepsilon_0\hbar}|\mu_{12}|^2\left(\Delta N^{(0)} - \frac{8\varepsilon_0 A\kappa\tau}{\hbar\omega}|\tilde{E}|^2\right).$$

By neglecting the term $d^2\tilde{E}/dt^2$ and $(d\tilde{E}/dt)^2$, when (9.28) holds, and substituting (9.21) for $\omega = \omega_0$, this becomes

$$(\kappa + \gamma)\frac{d\tilde{E}}{dt} - \kappa\gamma\left(\frac{\Delta N^{(0)}}{\Delta N_{\text{th}}} - 1\right)\tilde{E} + \frac{4A\kappa\tau}{\hbar^2}|\mu_{12}|^2|\tilde{E}|^2\tilde{E} = 0. \tag{9.29}$$

It can be rewritten in the form

$$\frac{d\tilde{E}}{dt} + (L - G + S|\tilde{E}|^2)\tilde{E} = 0 \quad \text{where} \tag{9.30}$$

$$L = \frac{\kappa\gamma}{\kappa+\gamma}, \qquad G = \frac{\Delta N^{(0)}}{\Delta N_{\text{th}}}L, \qquad S = \frac{4A\kappa\tau}{\hbar^2(\kappa+\gamma)}|\mu_{12}|^2. \tag{9.31}$$

Equation (9.30) is the van der Pol equation used in laser theory. The coefficient L is the damping constant, G is the amplification coefficient per unit time, and S is the saturation coefficient.

The solution of (9.30), when $G > L$, increases as $|\tilde{E}| \propto \exp[(G-L)t]$ as long as $|\tilde{E}|^2$ is small; as $|\tilde{E}|^2$ becomes large, it approaches the constant value $(G-L)/S$. When $\omega = \omega_0$, the value $(G-L)/S$ coincides with (9.23 or 24). The dynamics of a laser when external light is injected, and the characteristics of the laser in the presence of noise can be discussed by adding a perturbation term expressing the contribution of the incident light or noise to the right-hand side of the equation. When the incident light is coherent, so that its optical field may be expressed as

$$\tilde{E}_i e^{i\omega_i t} + \text{c.c.} ,$$

(9.30) is modified to become

$$\frac{d\tilde{E}}{dt} + (L - G + S|\tilde{E}|^2)\,\tilde{E} = \kappa_i \tilde{E}_i \exp[i(\omega_i - \omega)t] \tag{9.32}$$

because (9.30) is a differential equation of the complex amplitude varying as $\exp(i\omega t)$ so that the perturbation term varies at the frequency $\omega_i - \omega$. Here κ_i is a coefficient representing the time rate at which the incident light is injected. From (9.32) we may discuss, for example, frequency locking or synchronization of the laser with the injected light when its frequency ω_i is close to the laser transition frequency ω_0, and the associated change in the amplitude of oscillation as the injection frequency is gradually varied.

9.3 Multimode Oscillation

The fundamental equations of the semiclassical theory of a multimode laser are given by (9.8, 11, 12) as stated before. We may discuss various problems in a multimode laser by using these three simultaneous equations. Above all, we shall discuss here two basic phenomena: one is the problem of mode competition between two modes of oscillation and the other is the generation of combination frequencies when the laser is oscillating in three modes. The problem of mode locking is discussed in the subsequent section.

9.3.1 Competition Between Two Modes of Oscillation

If a laser is oscillating in two modes, its electric field can be expressed as

$$E(\boldsymbol{r}, t) = \tilde{E}_1(t)\, U_1(\boldsymbol{r})\, e^{i\omega_1 t} + \tilde{E}_2(t)\, U_2(\boldsymbol{r})\, e^{i\omega_2 t} + \text{c.c.} \tag{9.33}$$

The differential equations for $\tilde{E}_1(t)$ and $\tilde{E}_2(t)$ can be obtained as follows. The van der Pol equation for each mode is modified to have an additional term, which is introduced by intuitive considerations, so as to express the cross effect between the two modes: an equivalent equation can be obtained from an approximation of the fundamental equations in Sect. 9.1 for two-mode oscillation [9.3, 4].

Since the damping constant L of the resonator and the amplification coefficient G of the laser are, in general, different for the two different modes of oscillation, we shall assign the subscripts 1 and 2 to distinguish them. The saturation coefficients are given by S_1 and S_2 for self-saturation, similar to the case of single-mode oscillation. In addition, a cross-saturation effect must be considered in the case of two-mode oscillation. Cross-saturation is the effect where the intensity of oscillation in mode 1 is reduced as the intensity in mode 2 increases, and vice versa. Such a cross-saturation effect can be treated with additional terms in the van der Pol equations for the oscillation amplitudes $\tilde{E}_1$ and $\tilde{E}_2$ of the respective modes:

$$\frac{d\tilde{E}_1}{dt} + (L_1 - G_1)\,\tilde{E}_1 + S_1|\tilde{E}_1|^2\,\tilde{E}_1 + C_{12}|\tilde{E}_2|^2\,\tilde{E}_1 = 0\ , \tag{9.34 a}$$

$$\frac{d\tilde{E}_2}{dt} + (L_2 - G_2)\,\tilde{E}_2 + S_2|\tilde{E}_2|^2\,\tilde{E}_2 + C_{21}|\tilde{E}_1|^2\,\tilde{E}_2 = 0\ , \tag{9.34 b}$$

where C_{12} and C_{21} are cross-saturation coefficients. The coefficients L, G, S, and C are dependent on the modes of the resonator, the properties of the laser medium, its configuration in the resonator, the working conditions, and the intensity of excitation.

In some cases the cross-saturation effect is weaker than the self-saturation effect, and in other cases it is stronger. The oscillation behavior for the respective cases differs considerably, as shown below. The case for $C_{12}C_{21} < S_1S_2$ is called weak coupling, while the case for $C_{12}C_{21} > S_1S_2$ is called strong coupling.

The intensities I_1 and I_2 of the two modes of oscillation are proportional to $|\tilde{E}_1|^2$ nd $|\tilde{E}_2|^2$, respectively. Thus, by multiplying (9.34 a) by $\tilde{E}_1^*$, we obtain

$$\frac{1}{2}\frac{dI_1}{dt} = (G_1 - L_1)\,I_1 - S_1I_1^2 - C_{12}I_1I_2\ . \tag{9.35 a}$$

Similarly from (9.34 b) we obtain

$$\frac{1}{2}\frac{dI_2}{dt} = (G_2 - L_2)\,I_2 - S_2I_2^2 - C_{21}I_1I_2\ . \tag{9.35 b}$$

These two equations for the steady state give

$$I_1 = 0 \quad \text{or} \quad S_1I_1 + C_{12}I_2 = G_1 - L_1\ , \tag{9.36 a}$$

$$I_2 = 0 \quad \text{or} \quad C_{21}I_1 + S_2I_2 = G_2 - L_2\ , \tag{9.36 b}$$

since $\dot{I}_1 = 0$ and $\dot{I}_2 = 0$, where the dot denotes the time derivative.

We draw a graph with I_1 as the abscissa and I_2 as the ordinate, as in Fig. 9.2. Then (9.36 a) gives the I_2 axis and the solid slant line, while (9.36 b) gives the I_1 axis and the broken slant line. It can be seen from (9.35 a) that $\dot{I}_1 = 0$ on the solid slant line, while $\dot{I}_1 < 0$ above the line on the right and $\dot{I}_1 > 0$ below it on the left. The condition for $\dot{I}_2$ is similar with regard to the broken line. Therefore, the time variations of I_1 and I_2 in the four domains partitioned by the two slant lines are as shown in Fig. 9.2. It is thus found

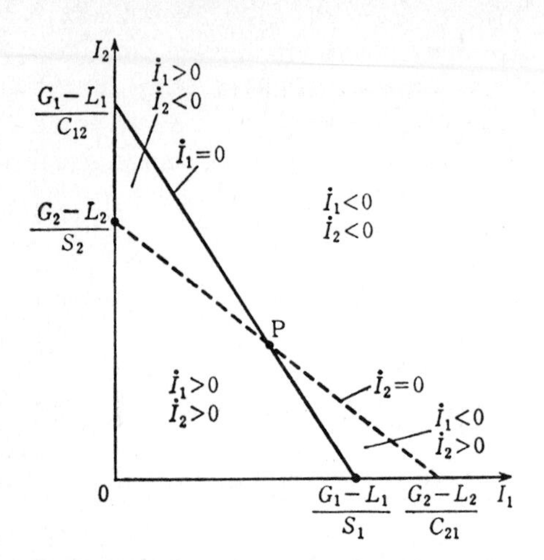

Fig. 9.2. Diagram for light intensities I_1 and I_2 of two-mode oscillation

that the states satisfying the steady-state conditions (9.36 a, b) simultaneously are given by the three black points in Fig. 9.2, of which the stable point is P. It should be noted that this holds for the weak-coupling case. When $C_{12} < S_1$, the solid slant line has a greater intercept on the I_2 axis than on the I_1 axis, while the broken line has the reverse. Figure 9.2 shows the case when the difference between G_1 and G_2 is not too large. If the difference is larger, the solid line and the broken line will not intersect at any positive values of I_1 and I_2, so that the laser oscillation will take place in the mode having a higher gain G than the other[2].

Leaving the strong-coupling case for later discussion, we discuss further the weak-coupling case in Fig. 9.2. The results of numerical calculation of the intensity variation of the two modes, I_1 and I_2, for various initial values are shown in Fig. 9.3. The arrow on each curve indicates the direction of variation with time. These curves are vertical when they cross the slant line for $\dot{I}_1 = 0$ corresponding to (9.36 a); they are horizontal when they cross the slant line for $\dot{I}_2 = 0$ corresponding to (9.36 b). Even without any numerical calculations, these considerations enable us to draw curves for variations of I_1 and I_2 which are more or less similar to those in Fig. 9.3 for the weak-coupling case in two-mode operation. Moreover, the directions of the arrows in Fig. 9.3 indicate that the two-mode oscillation is stable at the point P shown in Fig. 9.2.

In the strong-coupling case, when $C_{12}C_{21} > S_1S_2$, the broken line is steeper than the solid line in contrast to Fig. 9.2. Consequently, the signs of $\dot{I}_1$ and $\dot{I}_2$ in the triangular domains like those in Fig. 9.2, are opposite to the weak-coupling case. Therefore, the intersection of the two slant lines corresponding to (9.36 a, b) for strong coupling represents an unstable steady

2 Strictly speaking, the oscillating mode is not determined by G alone but rather by whether $(G - L)/S$ is greater or smaller.

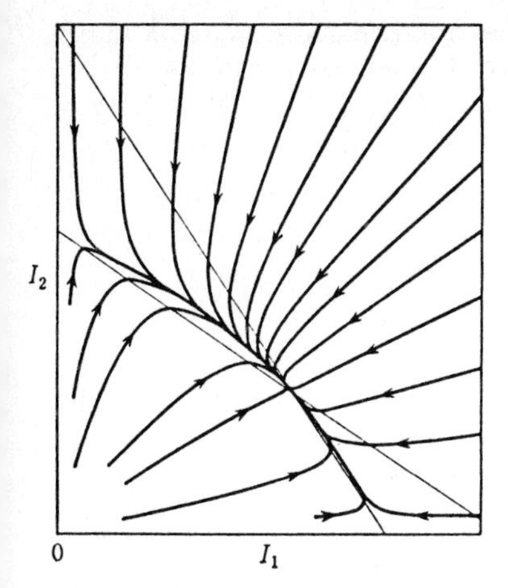

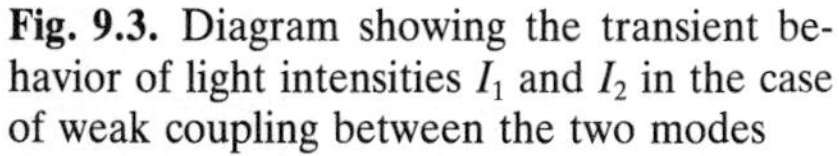
Fig. 9.3. Diagram showing the transient behavior of light intensities I_1 and I_2 in the case of weak coupling between the two modes

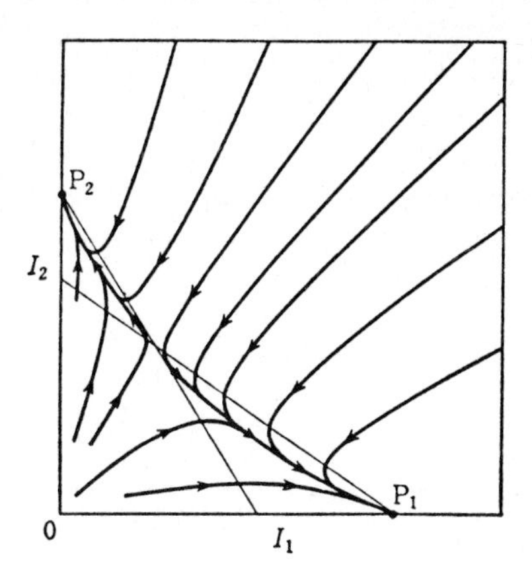

Fig. 9.4. Diagram showing the transient behavior of light intensities I_1 and I_2 in the strong-coupling case

state, while the intersections, P_1 and P_2, of the respective lines with $I_2 = 0$ and $I_1 = 0$ give the stable state. This becomes even more evident from the direction of the arrows in Fig. 9.4, where the results of numerical calculations of I_1 and I_2 from various initial values are given for the strong-coupling case.

Thus, in the strong-coupling case, only one of the modes oscillates, the other mode being suppressed. In the resulting steady state, therefore, the laser does not oscillate simultaneously in two modes, but it oscillates in only one of the modes. At the stable point P_1 in Fig. 9.4 it oscillates in mode 1, while at P_2 it oscillates in mode 2. Which of the bistable states is reached is dependent on the initial conditions, as can be seen from the curves in Fig. 9.4. For example, as the frequency of the laser resonator is gradually raised and lowered, there appears hysteresis in the oscillation intensity of each mode, as shown in Fig. 9.5. Such behavior can be explained by using Fig. 9.4 to aid the analysis; the two slant lines move in accordance with the gain profile $G(\omega)$ as the frequency ω is varied, while the other coefficients, L, S, and C are unchanged. The transition from two-mode to single-mode

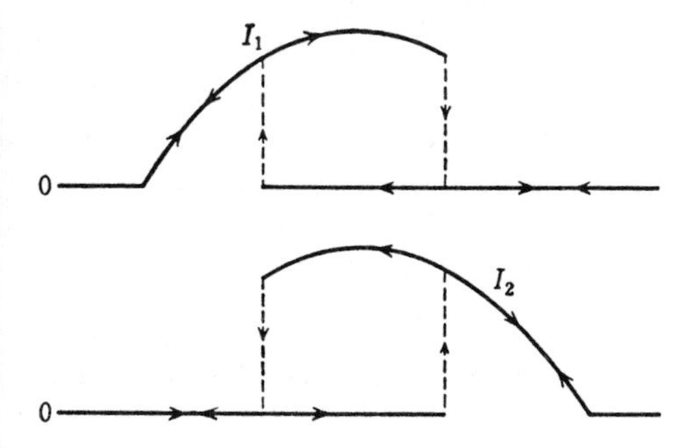

Fig. 9.5. Hysteresis of two-mode oscillation in the strong-coupling case

oscillation occurs when the intersection of the two slant lines reaches either the I_1 axis or the I_2 axis.

The frequencies, and also the polarizations, of the oscillations of the two modes mutually influence each other. The frequency separation between the two oscillations is generally wider than the mode spacing (known as mode pushing), and their optical polarizations tend to be linear and orthogonal to each other. We shall not, however, deal with such details here.

9.3.2 Existence of Combination Frequencies

As an example of multimode operation we shall consider oscillations in three modes at frequencies ω_1, ω_2, and ω_3, in increasing magnitude. Since the frequency separations between the oscillating modes are not exactly equal, in general, let us assume that $\omega_3 - \omega_2 < \omega_2 - \omega_1$ as in Fig. 9.6. Then the electric field $E(\boldsymbol{r}, t)$ and the polarization of the medium $P(\boldsymbol{r}, t)$ have frequency components at ω_1, ω_2, and ω_3.

According to one of the fundamental equations, (9.12), of a multimode laser, the time variation of the population inversion ΔN has a factor $\exp[i(\omega_k - \omega_j)t]$, so that ΔN fluctuates at the frequency $\omega_k - \omega_j$. This is called the pulsation of population inversion. The pulsation frequencies of ΔN in three-mode operation are $\omega_3 - \omega_2$ and $\omega_2 - \omega_1$. Since $\Delta N_{nm}\tilde{E}_m$ in the right-hand side of (9.11), which describes the time variation of the polarization, fluctuates at $(\omega_k - \omega_j) + \omega_m$, the induced polarization will fluctuate at the same frequency. This results in the electric field having the same frequency component, according to (9.8).

Mode indices k, j, m may take any values 1, 2, 3, ... for multimode operation in general. Now, if the frequency $\omega_k - \omega_j + \omega_m$ is different from

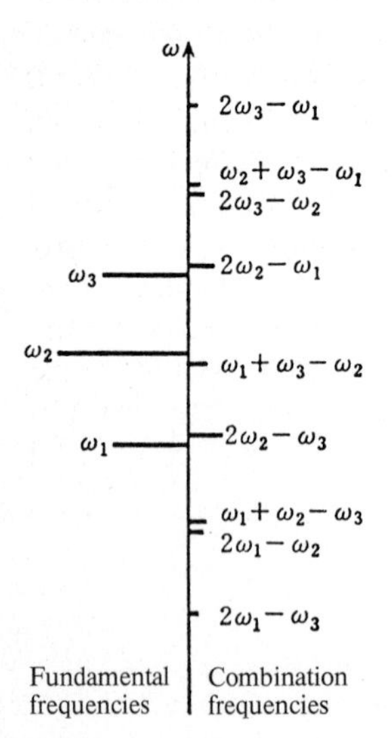

Fig. 9.6. Combination frequencies formed from the fundamental frequencies ω_1, ω_2, and ω_3 in three-mode oscillation

$\omega_1, \omega_2, \omega_3, \ldots$, it is called a combination frequency, while the original frequencies $\omega_1, \omega_2, \omega_3, \ldots$ are called fundamental frequencies. They are sometimes called a combination tone and fundamental tones, respectively.

In the case of three-mode oscillation, with $\omega_1 < \omega_2 < \omega_3$, there are nine combination frequencies, $2\omega_1 - \omega_3$, $2\omega_1 - \omega_2$, $\omega_1 + \omega_2 - \omega_3, \ldots$ in increasing order, as shown in Fig. 9.6. Though we considered three modes of oscillation initially, we must now treat a multimode oscillation with nine combination frequencies in the induced polarization, which produce an electric field for each of the nine frequencies. Similarly, in the case of two-mode oscillation, the electric fields at frequencies ω_1 and ω_2 will give a population pulsation at the frequency $\omega_2 - \omega_1$, which will give rise to electric fields at combination frequencies $2\omega_1 - \omega_2$ and $2\omega_2 - \omega_1$.

In fact, this is not all. When the combination frequency $2\omega_1 - \omega_2$ appears in addition to the two fundamental frequencies, a further pulsation component at $2\omega_1 - 2\omega_2$ will be developed in the population inversion, giving rise to polarization and electric field components at $3\omega_1 - 2\omega_2$ and $3\omega_2 - 2\omega_1$. They will be followed by an infinite number of successive combination frequencies at equal intervals of $\omega_2 - \omega_1$. In the three-mode case considered initially, there actually appear an infinite number of complicated higher-order combination frequencies. The nominal two- or three-mode oscillation means that the amplitudes of oscillation of two or three of the modes are much larger than those of the other modes, although there are actually an infinite number of oscillating modes.

In other words, there exists a solution satisfying the fundamental equations (9.8, 11, 12) for single-mode oscillation, whereas there is no solution for a finite number, two or three, of oscillating modes. Thus, except for single-mode oscillation, we must treat the coherent optical fields of an infinite number of modes. Since the electric field component at a combination frequency cannot fulfill the threshold condition by itself, it may sound reasonable not to designate it as an oscillation. It must be noted that the condition of oscillation in any one mode cannot be determined independently in the case of multimode operation. Consider two modes with either weak or strong coupling in which the two characteristic slant lines in Figs. 9.3 or 9.4 do not intersect in the first quadrant. Then, a primitive reasoning would infer that both the two modes are above threshold; as a matter of fact a single-mode oscillation is established, the other mode being suppressed in the steady state in either case. Again in the case of weak coupling, where simultaneous oscillations of two modes are allowed, there appears an infinite number of weak oscillating modes in addition to the two modes. This may be considered as an effect of parametric oscillation.

9.4 Mode Locking

Frequency differences between the adjacent modes of a multimode laser oscillator are not equal in general. This is because the effect of nonlinear dispersion of the laser medium results in the frequencies of the oscillating modes being slightly different from the proper frequencies of the longitudinal modes of the Fabry-Perot resonator, which are equally spaced. Therefore, when the output of a multimode laser is detected, many beat notes of slightly different frequencies are observed. Now, it is possible to make the mode frequency differences equal if a nonlinear optical element is inserted into the laser resonator, or if a modulator is inserted to modulate the loss or the refractive index with an applied *rf* voltage at a frequency close to the beat frequencies. This is called mode locking. When modes are locked by applying an *rf* voltage externally, it is called forced mode locking. When modes are locked without any external signals, it is called passive mode locking. Since the laser medium itself is optically nonlinear, mode locking may sometimes be obtained even without any additional nonlinear optical element, when the pumping condition and the resonator are appropriately adjusted. This is called self mode locking.

Self mode locking occurs because of the mode pulling effect between the fundamental frequencies and the combination frequencies. It is seen in the frequency spectrum of a three-mode oscillation in Fig. 9.6 that the pairs ω_1 and $2\omega_2 - \omega_3$, ω_2 and $\omega_1 + \omega_3 - \omega_2$, and ω_3 and $2\omega_2 - \omega_1$, are close to each other. Therefore, when these combination frequencies approach respectively ω_1, ω_2, and ω_3 within a certain range, frequency pulling occurs to give $\omega_1 = 2\omega_2 - \omega_3$, $\omega_2 = \omega_1 + \omega_3 - \omega_2$, and $\omega_3 = 2\omega_2 - \omega_1$. Since these are all equivalent to $\omega_3 - \omega_2 = \omega_2 - \omega_1$, the beat notes between all adjacent modes including combination frequencies, theoretically infinite in number, are unified. However, when more than three modes are oscillating, they are not always simultaneously mode locked.

Passive mode locking is similar to self mode locking, except that now strong combination frequencies for mode locking are generated by using a nonlinear optical element. Thus the terminology is often confused. The strength of the combination frequency is dependent not only on the nonlinear optical coefficient of the modulator but also markedly on its position in the resonator. The reason for this will be left for the reader to consider.

There are two methods of forced mode locking: one is amplitude modulation and the other is frequency modulation by a modulator in the laser resonator. When the laser field of mode n, $E_n \cos \omega_n t$, is amplitude modulated at ω_{mod} with a degree of modulation M, it becomes

$$E_n(1 + M\cos\omega_{\mathrm{mod}} t)\cos\omega_n t = \frac{M}{2} E_n \cos(\omega_n - \omega_{\mathrm{mod}})t + E_n\cos\omega_n t + \frac{M}{2} E_n \cos(\omega_n + \omega_{\mathrm{mod}})t \,. \tag{9.37}$$

It can be seen that sidebands at frequencies $\omega_n - \omega_{\text{mod}}$ and $\omega_n + \omega_{\text{mod}}$ are generated. Mode locking occurs when the neighboring modes $n-1$ and $n+1$ are pulled by these sidebands. Since the phases of both sidebands are equal, as can be seen from (9.37), the phases of the adjacent modes becomes nearly equal when they are locked by the modulation frequency ω_{mod} which is made close to the mode spacing.

Then the output power of the mode-locked laser is in the form of a train of pulses with a period given by $2\pi/\omega_{\text{mod}}$. To simplify the caculation, let us suppose that $2N+1$ modes are locked, all with the same amplitude and phase. Then the amplitude of the laser oscillation is expressed as

$$E(t) = \sum_{n=-N}^{N} E_0 \cos(\omega_0 + n\omega_{\text{mod}})t\,. \tag{9.38}$$

This is calculated to be

$$E(t) = E_0 \frac{\sin\left[\left(N+\frac{1}{2}\right)\omega_{\text{mod}}t\right]}{\sin\left(\frac{\omega_{\text{mod}}}{2}t\right)} \cdot \cos\omega_0 t\,,$$

so that the time variation of the laser power is given by

$$I(t) = E_0^2 \frac{\sin^2\left[\left(N+\frac{1}{2}\right)\omega_{\text{mod}}t\right]}{\sin^2\left(\frac{\omega_{\text{mod}}}{2}t\right)}\,. \tag{9.39}$$

Equation (9.39) has a maximum value of $(2N+1)^2E_0^2$, with a pulse width of about $T/(2N+1)$, peaked at $t=0$ and periodic intervals of $T = 2\pi/\omega_{\text{mod}}$. As an example, the waveform of the light intensity for $N=4$ is shown in Fig. 9.7.

The oscillation amplitudes of the modes in an actual mode-locked laser are not equal; the amplitude is usually largest at the center and becomes gradually smaller on either side. Moreover, since the phases of the locked

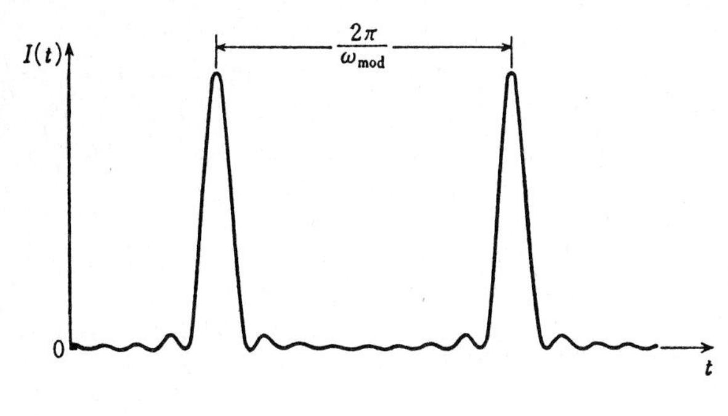

Fig. 9.7. Waveform of the mode-locked output for $N = 4$ from (9.39)

modes are often slightly shifted from one another, the profile of the pulse has larger wings and is a little broader than that of (9.39), the other features being similar. A first approximation to the pulse width is given by the reciprocal of the spectral range Δf of the mode frequencies of the multimode laser oscillation.

When the laser is mode locked by frequency modulation, a similar effect to amplitude modulation takes place to produce a pulse train, when the modulation frequency is close to the frequency of mode spacing. However, when the modulation frequency is somewhat different, another type of mode locking may occur where the amplitude of laser output does not change appreciably, but its phase is modulated at ω_{mod}. This is called FM mode locking. See [9.3, 4] or the original paper [9.5] for its analysis.

The reason why the output pulses of a mode-locked laser appear at intervals of $T = 2\pi/\omega_{\mathrm{mod}}$ is because a single pulse is circulating inside the laser resonator. In a Fabry-Perot resonator the time taken for light to make a round trip between the two mirrors shown in Fig. 9.8a is T. A ring resonator in which light circulates by reflection at three or four mirrors, shown in Fig. 9.8b, is resonant at frequencies such that the optical path length of the complete cycle L leads to a phase shift by an integral multiple of 2π. Then the resonant frequency ν_n of the longitudinal mode is given by

$$L = n\lambda = n\frac{c}{\nu_n}, \quad \text{so that } \nu_n = n\frac{c}{L} \tag{9.40}$$

where n is an integer. The frequency spacing of the modes is c/L and the period is $T = L/c$.

A laser with a ring resonator is called a ring laser. If a modulator is inserted into a ring laser, as shown in Fig. 9.8b, mode locking occurs when $2\pi/\omega_{\mathrm{mod}}$ is nearly equal to L/c, and output pulses are obtained at intervals of $T = 2\pi/\omega_{\mathrm{mod}}$. Whereas the light pulse in the Fabry-Perot laser travels back and forth through the laser medium and the modulator, it can be made to

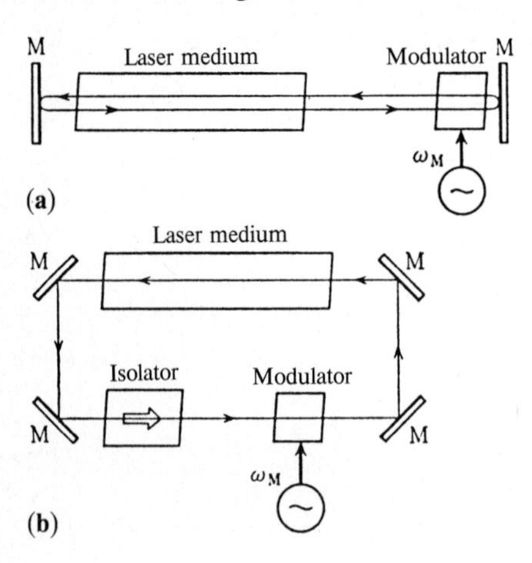

Fig. 9.8a, b. Mode-locked lasers: (**a**) a Fabry-Perot laser, and (**b**) a ring laser

travel in one direction in the ring laser by means of an optical isolator, so that the interaction between the laser field and atoms in every part of the resonator can be equivalent. Therefore, a detailed theoretical analysis of the ring laser is simpler, while superior performance is achieved experimentally.

The mechanism of mode locking in a ring laser can also be explained as below. If the absorption of the modulator in the ring laser fluctuates with period T, light passing through the modulator at the time when the absorption is strong will again be absorbed strongly by the modulator after a round trip and thus weakened successively. On the contrary, light passing through the modulator at the time when the absorption is weak will be sufficiently amplified by the laser medium since the loss is small in every cycle. Eventually the light energy is accumulated only at this phase, while circulating the ring as a lump of very intense light. Because a fraction of this light energy is transmitted through the mirror to the outside, the output of the mode-locked laser consists of a train of pulses with the period T.

Passive mode locking using a nonlinear absorber can be similarly explained by considering the temporal behavior of the laser. Here the modulator in Fig. 9.8 is replaced by a nonlinear absorber. The light intensity of the laser cannot be perfectly constant, and it always fluctuates by a small amount. When the light is stronger than average, the nonlinear absorber will be more saturated and the absorption coefficient will be smaller. Therefore, the stronger part of the fluctuating light will be strengthened more and more and the rest will be weakened more and more as they circulate in the resonator. Eventually, in the steady state, light of high intensity is lumped at one phase (sometimes two phases) during the cycle, so that the modes are locked. The pulse width of such a mode-locked laser depends not only on the gain bandwidth of the laser medium but also on its dispersion as well as the spectral and relaxation characteristics of the nonlinear absorber.

When the relaxation time of the laser medium is comparatively short (shorter than about $2\pi/\omega_{\mathrm{mod}}$), forced mode locking can be obtained by modulating the population inversion at frequency ω_{mod}, instead of using a modulator. This method is called synchronous pumping or gain modulation. Since there is no extra element in the resonator, and the bandwidth of the laser medium is wide, due to its short relaxation time, this method is favorable for obtaining very narrow mode-locked light pulses. A synchronously pumped dye laser can stably generate subpicosecond pulses (pulse width below 10^{-12} s) with a high repetition frequency.

9.5 Theory of the Gas Laser

The random motion of molecules (or atoms) in a gas gives rise to inhomogeneous broadening of the spectral line due to the Doppler effect, as described in Sect. 8.4. We consider a gas laser, consisting of two-level molecules with population inversion in a Fabry-Perot resonator. Each molecule in the gas laser is subjected to electric fields of two frequencies due to the Doppler effect, even when the laser is oscillating in a single mode. A theoretical description of gas lasers is, therefore, more complicated than one of ordinary lasers in which the atoms are at rest. It results in such characteristics as the Lamb dip, which does not appear in most other lasers.

The semiclassical theory of a gas laser was thoroughly studied using third-order perturbation by *W. E. Lamb* in 1962 and 1963 [9.6]. It was extended to include the effects of collisions in 1968 [9.7] and the large amplitude of oscillation in 1969 [9.8–10]. In addition, theories of multimode operation of a gas laser in a magnetic field (Zeeman laser) and a ring laser have been developed. But we restrict our discussions here to the fundamental characteristics of a single-mode gas laser.

9.5.1 Density Matrix of a Gas of Molecules in a Standing-Wave Field

Consider a single-mode planar standing wave between two plane-parallel mirrors, where the z axis is taken normal to the reflecting surface. Then the standing wave is composed of two plane waves progressing in the $+z$ and $-z$ directions, so that its real amplitude E may be expressed as

$$\begin{aligned} E(z,t) &= E\cos(\omega t + \varphi - kz) + E\cos(\omega t + \varphi + kz) \\ &= 2E\cos(\omega t + \varphi)\cos kz\ , \end{aligned} \tag{9.41}$$

where $k = \omega/c$ is equal to $n\pi/L$ (n is an integer and L is the distance between the two mirrors), as described in Sect. 3.4. In order to write the equation in a symmetrical form for convenience, the origin of the coordinate $z = 0$ has been taken at a crest of the standing wave.

The position of a moving molecule, which was at $z = z_0$ when $t = 0$, is given, prior to a following collision, as

$$z = z_0 + vt\ ,$$

where v is the z component of the molecular velocity. By substituting this equation into kz in (9.41), the electric field acting on the moving molecule can be expressed by

$$E^{(v)} = E\cos[(\omega - kv)t + \varphi - kz_0] + E\cos[(\omega + kv)t + \varphi + kz_0]\ , \tag{9.42}$$

which has two components of frequencies $\omega - kv$ and $\omega + kv$. Therefore, population pulsation at the frequency $2kv$ appears in the gas laser.

We have to calculate the polarization of a gas consisting of such molecules using the density matrix. Here we assume a model that represents actual molecules in the gas laser better than the two-level model by considering all excited levels, other than the two laser levels, and the ground level as a system of levels of a thermal reservoir. We further assume that the relaxation of the lower level 1 and the upper level 2 to the thermal reservoir is not dependent on the molecular population and that it can be given by the relaxation constants γ_1 and γ_2, respectively[3]. Unlike the system of two-level atoms discussed in Chap. 7, the sum $N_2 + N_1$ or $\varrho_{22} + \varrho_{11}$ of the populations in the upper and the lower levels is not constant but changes with time in this case.

The equation of motion of the density matrix (7.39) must then be modified to include the relaxation term, and the perturbation Hamiltonian $\mathcal{H}'$ of the dipole moment operator μ in the electric field (9.41) is assumed to be given by

$$\mathcal{H}' = -\mu E(z,t) . \tag{9.43}$$

In order to express the time derivative of the density matrix $d\varrho/dt$ of molecules moving in the electric field which varies in space as in (9.41), we must have

$$\frac{d}{dt} = \frac{\partial}{\partial t} + v\frac{\partial}{\partial z} ,$$

which is similar to the time derivative in the Euler equations of hydrodynamics. Thus the differential equations of the density-matrix elements for the two levels with the eigenfrequency ω_0 are expressed as

$$\left(\frac{\partial}{\partial t} + v\frac{\partial}{\partial z}\right)\varrho_{11} + \gamma_1(\varrho_{11} - \varrho_{11}^{(0)}) = \frac{\mathrm{i}}{\hbar}\mathcal{H}'_{21}\varrho_{12} + \text{c.c.} , \tag{9.44 a}$$

$$\left(\frac{\partial}{\partial t} + v\frac{\partial}{\partial z}\right)\varrho_{22} + \gamma_2(\varrho_{22} - \varrho_{22}^{(0)}) = -\frac{\mathrm{i}}{\hbar}\mathcal{H}'_{21}\varrho_{12} + \text{c.c.} , \tag{9.44 b}$$

$$\left(\frac{\partial}{\partial t} + v\frac{\partial}{\partial z} - \mathrm{i}\omega_0\right)\varrho_{12} + \gamma\varrho_{12} = -\frac{\mathrm{i}}{\hbar}\mathcal{H}'_{12}(\varrho_{22} - \varrho_{11}) . \tag{9.44 c}$$

Here γ is the transverse relaxation constant. It is given by $\gamma = (\gamma_1 + \gamma_2)/2$ if phase-changing collisions between molecules are negligible, but it is otherwise usually given as $\gamma > (\gamma_1 + \gamma_2)/2$ in an actual gas. The quantities $\varrho_{11}^{(0)}$ and $\varrho_{22}^{(0)}$ are the stationary values of ϱ_{11} and ϱ_{22} in the absence of perturbation from the optical field. They give the population inversion as

$$\Delta\varrho^{(0)} = \varrho_{22}^{(0)} - \varrho_{11}^{(0)} > 0 . \tag{9.45}$$

3 In Sect. 8.2 we discussed the saturated absorption of such a quasi-two-level atom, using a rate-equation approximation.

Since molecules have a velocity distribution, let us first calculate the density matrix for a group of molecules with a velocity component between v and $v + dv$. Then we integrate it over the entire molecular velocity distribution to obtain the polarization and the population inversion of the gas. In a frame of reference moving with a molecule at $z = z_0 + vt$, the molecule appears to be at rest, and the equations of motion of the density matrix (9.44), as expressed in the rest frame, are simplified to the form

$$\left(\frac{\partial}{\partial t}+\gamma_1\right)(\varrho_{11}-\varrho_{11}^{(0)}) = \mathrm{i}V^*\varrho_{12} + \mathrm{c.c.}\ , \tag{9.46 a}$$

$$\left(\frac{\partial}{\partial t}+\gamma_2\right)(\varrho_{22}-\varrho_{22}^{(0)}) = -\mathrm{i}V^*\varrho_{12} + \mathrm{c.c.}\ , \tag{9.46 b}$$

$$\left(\frac{\partial}{\partial t}+\gamma-\mathrm{i}\omega_0\right)\varrho_{12} = -\mathrm{i}V(\varrho_{22}-\varrho_{11})\ . \tag{9.46 c}$$

Here $\hbar V$ is the perturbation as seen in the moving frame, given by

$$\hbar V = -\mu_{12}E^{(v)}\ ,$$

since the field acting on the molecule is $E^{(v)}$ from (9.42). Thus, by putting

$$x = \frac{\mu_{12}E}{\hbar}\,\mathrm{e}^{\mathrm{i}\varphi}\ , \qquad \phi = kvt + kz_0\ , \tag{9.47}$$

we have

$$V = -\cos\phi\,(x\,\mathrm{e}^{\mathrm{i}\omega t} + \mathrm{c.c.}) \tag{9.48}$$

from (9.42).

9.5.2 Approximate Solutions by Iteration

Even with the above approach, (9.46) cannot be solved analytically, since ϕ is a function of time as shown by (9.47), and V has two frequency components. But easier calculations can be made by successive approximation or iteration. As shown previously, ϱ_{12} can be expressed in odd-order terms of x, and ϱ_{11} and ϱ_{22} can be expressed in even-order terms, so that we may write

$$\varrho_{12} = \varrho_{12}^{(1)}x + \varrho_{12}^{(3)}|x|^2x + \dots\ , \tag{9.49 a}$$

$$\varrho_{11} = \varrho_{11}^{(0)} + \varrho_{11}^{(2)}|x|^2 + \dots\ , \tag{9.49 b}$$

$$\varrho_{22} = \varrho_{22}^{(0)} + \varrho_{22}^{(2)}|x|^2 + \dots\ . \tag{9.49 c}$$

Considering only the steady state, we may obtain the solution of (9.46) by substituting the terms of (9.49) successively from the lowest-order term.

To begin with, in the first-order approximation the second- and higher-order terms are neglected, so that the right-hand side of (9.46 c) becomes $-\mathrm{i}V\Delta\varrho^{(0)}$, and we obtain

$$\varrho_{12}^{(1)} = \frac{\mathrm{i}}{2}\Delta\varrho^{(0)}\,\mathrm{e}^{\mathrm{i}\omega t}\left(\frac{\mathrm{e}^{\mathrm{i}\phi}}{\gamma + \mathrm{i}(\omega - \omega_0 + kv)} + \frac{\mathrm{e}^{-\mathrm{i}\phi}}{\gamma + \mathrm{i}(\omega - \omega_0 - kv)}\right) \tag{9.50}$$

by using (9.47, 48). Then, the second-order approximation of ϱ_{11} can be obtained, using the first-order approximation (9.50) for ϱ_{12} in (9.46 a):

$$\varrho_{11}^{(2)} = \frac{\Delta\varrho^{(0)}}{4\gamma_1}\left(\frac{1 - \dfrac{\mathrm{i}\gamma_1}{2kv}\mathrm{e}^{2\mathrm{i}\phi}}{\gamma + \mathrm{i}(\omega - \omega_0 + kv)} + \frac{1 + \dfrac{\mathrm{i}\gamma_1}{2kv}\mathrm{e}^{-2\mathrm{i}\phi}}{\gamma + \mathrm{i}(\omega - \omega_0 - kv)}\right) + \text{c.c.} \tag{9.51}$$

Similarly, $\varrho_{22}^{(2)}$ can be calculated from (9.46 b), with the same result, except that the sign of (9.51) is changed and γ_1 is replaced by γ_2. It should be noted that in (9.51) the molecular population pulsates at the frequency $2kv$, since ϕ varies in time as kvt. This is the population pulsation. Returning to the rest frame, we have $\phi = kz$, so that the population inversion of molecules with a velocity distribution varies more or less sinusoidally in the z direction with a period $2\pi/2k = \lambda/2$, where λ is the wavelength. Since the distance between crests of the standing wave is one half-wavelength as seen in Fig. 9.9, we have a corresponding distribution of population inversion. This is known as the spatial hole-burning effect.

In the case of multimode oscillation, spatial hole-burning with a long period, corresponding to the wavelength of the beat frequency between the modes, appears in addition. In the case of single-mode operation, on the other hand, the hole-burning effect may be neglected for ordinary gas lasers, since it has only a very short period. Therefore, the second-order approximation for population inversion $\Delta\varrho^{(2)} = \varrho_{22}^{(2)} - \varrho_{11}^{(2)}$ becomes

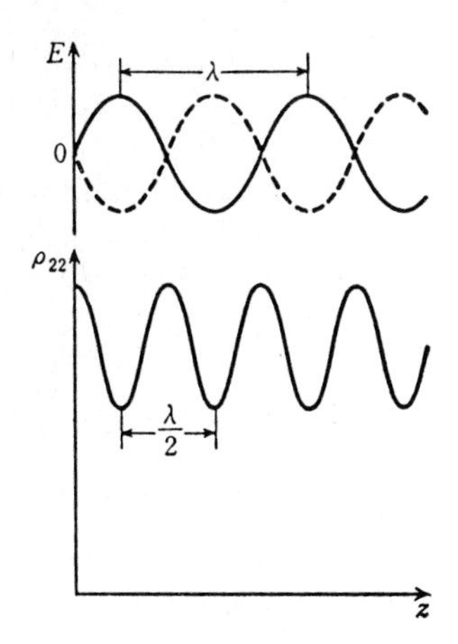

Fig. 9.9. Spatial hole-burning due to standing waves

$$\Delta\varrho^{(2)} = -\tau\Delta\varrho^{(0)}\left(\frac{\gamma}{\gamma^2 + (\omega - \omega_0 + kv)^2} + \frac{\gamma}{\gamma^2 + (\omega - \omega_0 - kv)^2}\right), \quad (9.52)$$

where τ is the longitudinal relaxation time given by

$$\tau = \frac{1}{2}\left(\frac{1}{\gamma_1} + \frac{1}{\gamma_2}\right). \quad (9.53)$$

Equation (9.52) shows that the molecular velocity distribution decreases for the upper level and increases for the lower level at the velocity

$$v = \pm\frac{\omega - \omega_0}{k}$$

where ω is the frequency of the laser. Thus the hole-burning effect on population inversion is similar to that in Fig. 8.7, with holes appearing on both sides of the velocity component, where N_1 and N_2 are interchanged for inverted population.

The third-order approximation is calculated by substituting $\varrho_{22} - \varrho_{11} = \Delta\varrho^{(0)} + \Delta\varrho^{(2)}|x|^2$ into the right-hand side of (9.46c) to give

$$\varrho_{12}^{(3)} = \frac{i}{2}\Delta\varrho^{(2)}\,e^{i\omega t}\left(\frac{e^{i\phi}}{\gamma + i(\omega - \omega_0 + kv)} + \frac{e^{-i\phi}}{\gamma + i(\omega - \omega_0 - kv)}\right). \quad (9.54)$$

The final expression can then be obtained by substituting (9.52) into the above. Solutions of higher-order approximation may be calculated by repeating such an iteration procedure.

Since $\varrho_{12}(v)$ is obtained in this manner with the velocity component v as a parameter, the polarization of the gas can be calculated by integrating $\varrho_{12}(v)\mu_{21}$ + c.c. with respect to the velocity and spatial distributions of the molecules. This calculation, however, is not easy. Instead, we shall first obtain the nonlinear subsceptibility $\chi(v)$ only for those molecules with velocities in the range between v and $v + dv$, and then integrate it with respect to the velocity distribution. Now we expand $\chi(v)$ in powers of $E^2 = \hbar^2|x|^2/|\mu_{12}|^2$ in the form

$$\chi(v) = \chi^{(1)}(v) + \chi^{(3)}(v)E^2 + \dots . \quad (9.55)$$

Then the linear susceptibility is obtained from (9.50) as

$$\chi^{(1)}(v) = \frac{i}{2\varepsilon_0\hbar}|\mu_{12}|^2 N(v)\Delta\varrho^{(0)}$$

$$\times\left(\frac{1}{\gamma + i(\omega - \omega_0 + kv)} + \frac{1}{\gamma + i(\omega - \omega_0 - kv)}\right). \quad (9.56)$$

The third-order nonlinear susceptibility is calculated from (9.54, 52) as

$$\chi^{(3)}(v) = -\frac{\mathrm{i}\tau}{\varepsilon_0 \hbar^3} |\mu_{12}|^4 N(v) \Delta\varrho^{(0)}$$
$$\times \left(\frac{\gamma}{\gamma^2 + (\omega - \omega_0 + kv)^2} + \frac{\gamma}{\gamma^2 + (\omega - \omega_0 - kv)^2} \right)$$
$$\times \left(\frac{1}{\gamma + \mathrm{i}(\omega - \omega_0 + kv)} + \frac{1}{\gamma + \mathrm{i}(\omega - \omega_0 - kv)} \right) . \quad (9.57)$$

Higher-order susceptibilities can be obtained by repeating this iteration procedure.

When the molecular velocity in the gas obeys the Maxwell-Boltzmann distribution, the linear and the third-order susceptibilities of the gas can be obtained from integration of (9.56 and 57), respectively, using (8.36). With the use of the plasma dispersion function (Sect. 8.4), the result of integrating (9.56) can be written as

$$\chi^{(1)} = \int_{-\infty}^{\infty} \chi^{(1)}(v)\, dv = -\frac{N\Delta\varrho^{(0)}}{\varepsilon_0 \hbar k u} |\mu_{12}|^2 Z(\zeta) ,$$
$$\zeta = \frac{\omega - \omega_0 - \mathrm{i}\gamma}{ku} . \quad (9.58)$$

Since the calculation of higher orders is lengthy in general [9.10, 11], we write only $\chi^{(1)}$ and $\chi^{(3)}$ in the Doppler limit approximation. When the Doppler width is wide in comparison with the homogeneous width, $\gamma \ll ku$, by using the real variable $\xi = (\omega - \omega_0)/ku$, we obtain

$$\chi^{(1)} = -\frac{\sqrt{\pi}\, N\Delta\varrho^{(0)}}{\varepsilon_0 \hbar k u} |\mu_{12}|^2 \mathrm{e}^{-\xi^2} \left(\frac{2}{\sqrt{\pi}} \int_0^{\xi} \mathrm{e}^{x^2} dx - \mathrm{i} \right) . \quad (9.59)$$

The third-order susceptibility is likewise obtained to be

$$\chi^{(3)} = \frac{\sqrt{\pi}\, N\Delta\varrho^{(0)}\tau}{\varepsilon_0 \hbar^3 k u} |\mu_{12}|^4 \mathrm{e}^{-\xi^2} \left(\frac{\mathrm{i}}{\gamma} + \frac{1}{\omega - \omega_0 - \mathrm{i}\gamma} \right) . \quad (9.60)$$

9.5.3 Output Characteristics in the Third-Order Approximation (Lamb Dip)

When the nonlinear susceptibility of a gas with population inversion is known, the amplitude and the frequency of single-mode oscillation can be determined by the use of (9.17). For the sake of simplicity, we assume that the gaseous medium is homogeneous throughout the interior of the laser resonator[4]. Then the relation between the amplitude $\tilde{E}$ of the oscillating

4 Since the actual laser medium is confined in a laser tube and the excitation is not at all homogeneous, $\tilde{P}$ is expressed by the inverse transform of (9.2) weighted with the mode function. Therefore χ is expressed by a similar weighted average.

mode and the polarization $\tilde{P}$ of that mode can be written, using the nonlinear susceptibility χ calculated above, as

$$\tilde{P} = \varepsilon_0 \chi \tilde{E} . \tag{9.61}$$

When the mode function is

$$U(\boldsymbol{r}) = \cos kz , \tag{9.62}$$

the optical field of the plane standing wave (9.41) gives

$$\tilde{E} = E\,\mathrm{e}^{\mathrm{i}\varphi} , \quad \text{and} \quad x = \frac{\mu_{12}\tilde{E}}{\hbar}$$

from (9.47). The amplitude can then be determined from (9.17 a) by using the imaginary part of

$$\chi = \chi^{(1)} + \chi^{(3)} E^2 + \dots , \tag{9.63}$$

while the frequency of oscillation can be determined from (9.17 b) by using the real part of (9.63). Since (9.17 a) for the steady state becomes

$$\kappa + (\omega/2)\,(\chi^{(1)''} + \chi^{(3)''} E^2 + \dots) = 0 ,$$

the third-order approximation gives

$$E^2 = \left(-\chi^{(1)''} - \frac{2\kappa}{\omega}\right) \Big/ \chi^{(3)''} . \tag{9.64}$$

The imaginary part of (9.59) is substituted into this to calculate the threshold value of population inversion at which E^2 appears. The result for $\omega = \omega_0$ is

$$\Delta N_{\mathrm{th}} = (N\Delta\varrho^{(0)})_{\mathrm{th}} = \frac{2\,\varepsilon_0 \hbar \kappa u}{\sqrt{\pi}\,|\mu_{12}|^2 c} . \tag{9.65}$$

The reason for the discrepancy by a factor $ku/\sqrt{\pi}\gamma$ with (9.21) for $\omega = \omega_0$ is because the line profile is Gaussian instead of Lorentzian.

Substituting the imaginary parts of (9.59, 60) into (9.64), we obtain the intensity of oscillation:

$$E^2 = \frac{\hbar^2 \gamma}{|\mu_{12}|^2 \tau} \cdot \frac{1 - \dfrac{1}{\mathcal{N}} \exp\left(\dfrac{\omega - \omega_0}{ku}\right)^2}{1 + \dfrac{\gamma^2}{(\omega - \omega_0)^2 + \gamma^2}} , \quad \text{where} \tag{9.66}$$

$$\mathcal{N} = \frac{N\Delta\varrho^{(0)}}{\Delta N_{\mathrm{th}}}$$

is the relative intensity of excitation.

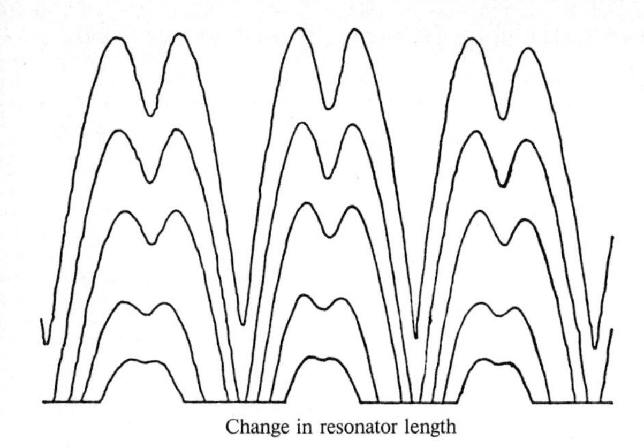

Fig. 9.10. Lamb dips observed in the 1.15 μm He-Ne laser for different excitation strengths

The above approximation is for the case when the Doppler broadening ku is large and the homogeneous broadening γ is small. The numerator of (9.66) exhibits a broad peak as a function of the frequency, whereas the denominator shows a sharp increase within the homogeneous width centered at ω_0. Then, the oscillation intensity of a gas laser decreases near the central frequency of oscillation, as observed in Fig. 9.10. The laser output appears every half wavelength as the frequency is scanned by changing the length of the resonator. Such a dip in the output characteristic was foreseen by *W. E. Lamb* in 1962 in his theory of the gas laser, and verified experimentally shortly afterward. Thus it is called the Lamb dip.

The frequency of steady-state oscillation can be obtained from (9.17 b) by putting $d\phi/dt = 0$. Substituting the real part of (9.59) and using (9.65), we obtain the first-order approximation which gives the frequency-pulling effect similar to (9.20) or (5.53), except that γ in these equations is replaced by $(\sqrt{\pi}/2)\,ku$, which is equal to 1.066 times the HWHM. Since the third-order calculation of χ' gives rise to a decrease of pulling in proportion to E^2, the laser frequency departs from the proper frequency of the resonator as the intensity of oscillation increases. This is called the frequency-pushing effect. The frequency-pushing effect is also observed in multimode oscillations, where the neighboring modes tend to push each other so as to increase the frequency separation between the modes, as the intensity of oscillation is increased.

9.6 Fully Quantum-Mechanical Theory of the Laser

The fluctuation of the electromagnetic field of light is neglected in the semiclassical theory of the laser which has been discussed so far. Therefore the semiclassical theory is not capable of discussing fluctuations of the amplitude of oscillation, namely, laser noise and the spectral width of the laser radiation. Moreover, it is not possible to discuss photon statistics and higher-order correlations of the laser radiation since the electromagnetic field is treated classically without using quantization.

In order to discuss the effects of the so-called quantum noise, it is necessary to treat both the electromagnetic field and the atoms quantum-mechanically. In the model shown in Fig. 9.1 we have to consider mutual interactions of the laser medium with the pumping source and with the cold bath, and the coupling of the laser with the external world. These must all be treated quantum-mechanically, including spontaneous emission and zero-point vibration. Quantum-mechanical treatments of the laser as a nonlinear, non-equilibrium and open system were developed by *H. Haken*, *W. E. Lamb* and others towards the latter half of 1960. Leaving the details to [9.3, 4, 12, 13], here we only briefly review these quantum-mechanical theories of the laser.

The optical field interacting with an atom is expressed as an operator acting on the state functions of the electromagnetic field. When the electromagnetic field is expanded into modes, the field operator is expressed as creation and annihilation operators of photons in each mode. It should be remarked that in the laser theory, where not only the atoms but also the optical field is quantized, the numbers of atoms and photons assume much greater values than 1. Besides, the system of atoms and photons is in contact with the thermal reservoir through losses and the output coupling of the resonator, while the atoms in the laser medium are pumped for inverted population.

It is even more difficult in the fully quantum-mechanical theory than in the semiclassical theory of the laser to discuss the Hamiltonian of the entire system, including the effects of the thermal reservoir and the process of excitation. Moreover, since the optical field operator which represents the amplitude of the laser oscillation cannot describe noise by a time-varying continuous function as in the semiclassical theory, fluctuations in the fully quantum-mechanical theory do not immediately correspond to the observed noise or the spectral width of the laser output. Therefore, in order to overcome this deficiency, various models and approximations have been used in the quantum-mechanical theories of the laser.

The first of these is the theory which treats the laser with a classical Langevin equation. Here fluctuations due to the interaction of atoms in the laser medium with the thermal reservoir and the excitation source are treated in the form of a phenomenological Langevin force, namely, as a random disturbance. The quantum-mechanical Heisenberg equation is solved by treating the Langevin force as a perturbation. The laser theory based on this method was developed by *H. Haken* and his collaborators in 1966.

The second is a laser theory in which the Schrödinger representation is used to express the density matrix in general, by using the base vectors of the photon-number states in the absence of a thermal reservoir. The effect of the thermal reservoir is treated as a perturbation. This method is easy to handle since the electromagnetic field is expanded in terms of the photon-number states as with most other problems of quantum mechanics. On the other hand, it is difficult to treat the coherence of laser light, which is given by the off-diagonal matrix elements. *M. O. Scully* and *W. E. Lamb*, in 1967, obtained the spectral width of laser oscillation from the density-matrix formalism based on the photon-number states. Subsequently, it has been further developed to discuss many other problems of the laser.

The third method uses the so-called Fokker-Planck equation. *H. Risken*, in 1965, semiclassically interpreted the fully quantum mechanical laser theory of *H. Haken* and others, and rewrote the distribution function of the electromagnetic field of the laser in the form of the Fokker-Planck equation. Moreover, by using the expression for the quantum-mechanical coherent state introduced by *R. J. Glauber*, in 1963, *H. Haken* and many others have developed quantum-mechanical laser theories employing the Fokker-Planck equation [9.14, 15].

These fully quantum-mechanical laser theories were used to calculate the distribution of laser-photon counts to be compared with experimental results. Further, from the photon counting statistics, higher-order moments of quantum noise and higher-order coherence have been investigated both theoretically and experimentally.

Problems

9.1 In obtaining (9.7) using the rotating-wave approximation and the slowly varying envelope approximation, the term $d\tilde{P}_n/dt$ has been neglected. Show that it is a second-order small quantity on the assumptions adopted here.

Answer: Because $d\tilde{P}_n/dt \cong \gamma\tilde{P}_n \ll \omega_n\tilde{P}_n$ and $\tilde{P}_n \ll \varepsilon_0\tilde{E}_n$.

9.2 The frequency of the laser has been calculated to be given by (9.20), where Lorentzian profiles of the optical resonator and the spectral line have been implicitly assumed. Explain how they are assumed in Sect. 9.1.

Answer: Exponential damping of the resonator has been assumed in the use of the damping constant κ_n. Equation (9.9) has been rewritten by using (7.49) which assumes a constant relaxation rate γ of the polarization. The power spectrum of a damped oscillation takes a Lorentzian profile.

9.3 Derive formulas for the frequency and the threshold population inversion of steady-state oscillation in a single-mode laser, using (9.17 a, b) instead of (9.14 a, b).

Hint: Put $d/dt = 0$ in (9.17 a, b) and use (8.31) for $|x|^2 = 0$.

9.4 Find a formula that describes the build up of the laser intensity $W = |\tilde{E}|^2$ by using the van der Pol equation with the initial value W_0 at $t = 0$.

Answer: The result from (9.30) is identical to (6.24) if we take $a - \kappa = G - L$ and $ab = S$ or $W_{ss} = (G - L)/S$.

9.5 The frequency and amplitude of laser oscillation are perturbed with the injected monochromatic light as given by (9.32). If the laser field takes a stationary amplitude, it may be written in the form

$$E(t) = a \exp[\mathrm{i}\omega_L + \mathrm{i}\theta(t)] + \mathrm{c.c.}$$

so that we have

$$\tilde{E}(t) = a \exp[\mathrm{i}(\omega_L - \omega)t + \mathrm{i}\theta(t)] \; .$$

Show that the solution of (9.32) in this case can be expressed in the form

$$t - t_0 = \int_{\varphi_0}^{\varphi} \frac{d\varphi}{\omega_i - \omega - b \sin\varphi} \; ,$$

where $\varphi(t) = (\omega_i - \omega_L)t - \theta(t)$, and $b = \kappa_i \tilde{E}_i / a$.

Hint: $d\theta(t)/dt = \omega - \omega_L + b \sin\varphi(t)$.

9.6 The cross-saturation parameter in (9.34) for a gas laser can be approximately given by

$$C_{12} = S_1 \gamma^2 [(\omega_1 - \omega_2)^2 + \gamma^2]^{-1} \; .$$

Show that, with proper choice of absorber gas at low pressure inside the resonator, the mode coupling in a gas laser can be made to be strong, although it is weak in the absence of the absorber.

Answer: The coefficients in (9.34) are now written as $G = G_g - G_a$, $S = S_g - S_a$, and $C = C_g - C_a$, where subscripts g and a denote the gain medium and the absorber, respectively. From (9.31), G_a/S_a is seen to be proportional to γ/τ or p_a^2, where p_a is the gas pressure. If p_a is so chosen that $S_a \approx S_g$, we have $S \approx 0$, $G_a \ll G_g \approx G$, and $C_a \ll C_g \approx C$, resulting in strong coupling.

9.7 Insertion of a few-mm-thick glass plate into the resonator of a mode-locked laser generating subpicosecond pulses gives rise to considerable broadening of the pulse. Discuss the pulse-broadening effect due to normal dispersion of the glass plate. The refractive index of BK 7 glass, for example, is $\eta = 1.5143$ at 656 nm and 1.5168 at 588 nm.

Answer: A subpicosecond pulse has a frequency spread of the order of 10^{13} Hz, while normal dispersion of BK7 is $d\eta/d\nu \cong 5 \times 10^{-17}\ \mathrm{Hz}^{-1}$. Then the difference in transit times of two components with a frequency difference of 10^{13} Hz through a glass plate of 3 mm thickness will be 0.05 ps. Because of the multiple transit of light pulses in the laser resonator, the resulting broadening can be more than ten times this.

9.8 Spatial hole burning on the molecular population in a gas laser as shown in Fig. 9.9 will be shallower at higher temperatures if other parameters are unchanged. Explain the reason for this.

Answer: Equation (9.51) gives the spatial hole-burning effect for molecules having the velocity component v. The hole-burning effect integrated over the whole velocity distribution is smeared more strongly at higher temperatures.

9.9 Fluctuations in the laser can be treated by using a van der Pol equation with a random perturbation term in the form

$$\frac{d\tilde{E}}{dt} + (L - G + S|\tilde{E}|^2)\,\tilde{E} = f(t)\ ,$$

where $f(t)$ represents the time-dependent perturbation. Solve this equation on the assumption that the perturbation is weak.

Answer: $\tilde{E}(t) = E_0 + \exp[-2(G-L)t] \int_0^t f(t) \exp[2(G-L)t]\,dt$

$+ [\tilde{E}(0) - E_0] \exp[-2(G-L)t], \quad$ where $E_0^2 = (G-L)/S$.

References

Chapter 1

1.1 A. L. Schawlow, C. H. Townes: Phys. Rev. **112**, 1940 (1958); H. Haken: *Laser Theory* (Springer, Berlin, Heidelberg 1984)
1.2 S. L. Shapiro (ed.): *Ultrashort Light Pulses*, Topics Appl. Phys., Vol. 18, 2nd ed. (Springer, Berlin, Heidelberg 1984)
1.3 W. Koechner: *Solid-State Laser Engineering*, Springer Ser. Opt. Sci., Vol. 1 (Springer, Berlin, Heidelberg 1976)
1.4 A. A. Kaminskii: *Laser Crystals*, Springer Ser. Opt. Sci., Vol. 14 (Springer, Berlin, Heidelberg 1981)
1.5 R. Beck, W. Englisch, K. Gürs: *Table of Laser Lines in Gases and Vapors*, Springer Ser. Opt. Sci., Vol. 2, 3rd ed. (Springer, Berlin, Heidelberg 1980)
1.6 H. E. White: *Introduction to Atomic Spectra* (McGraw-Hill, New York 1934); I. I. Sobelman: *Atomic Spectra and Radiative Transitions*, Springer Ser. Chem. Phys., Vol. 1 (Springer, Berlin, Heidelberg 1979)
1.7 G. Brederlow, E. Fill, K. J. Witte: *The High-Power Iodine Laser*, Springer Ser. Opt. Sci., Vol. 34 (Springer, Berlin, Heidelberg 1983)
1.8 C. K. Rhodes (ed.): *Excimer Lasers*, Topics Appl. Phys., Vol. 30, 2nd ed. (Springer, Berlin, Heidelberg 1984)
1.9 F. P. Schäfer (ed.): *Dye Lasers*, Topics Appl. Phys., Vol. 1, 2nd ed. (Springer, Berlin, Heidelberg 1977)
1.10 M. Maeda: Rev. Laser Eng. **8**, 694, 803, 958 (1980); **9**, 85, 190 (1981) [in Japanese]; *Laser Dyes* (Academic, New York) to be published
1.11 H. Kressel (ed.): *Semiconductor Devices*, Topics Appl. Phys., Vol. 39, 2nd ed. (Springer, Berlin, Heidelberg 1982)
1.12 S. M. Sze: *Physics of Semiconductor Devices*, 2nd ed. (John Wiley, New York 1981); K. Seeger: *Semiconductor Physics*, Springer Ser. Solid-State Sci., Vol. 40, 2nd ed. (Springer, Berlin, Heidelberg 1982)

Chapter 2

2.1 M. Born, E. Wolf: *Principles of Optics*, 6th ed. (Pergamon, Oxford 1980); B. Saleh: *Photoelectron Statistics*, Springer Ser. Opt. Sci., Vol. 6 (Springer, Berlin, Heidelberg 1978)
2.2 D. Gabor: J. Inst. Electr. Eng. **93**, 429 (1946)
2.3 F. Zernike: Physica **5**, 785 (1938)

Chapter 3

3.1 H. K. V. Lotsch: Optik **32**, 116, 189, 299 and 553 (1970/71)
3.2 R. G. Hunsperger: *Integrated Optics: Theory and Technology*, Springer Ser. Opt. Sci., Vol. 33, 2nd ed. (Springer, Berlin, Heidelberg 1984)
3.3 A. B. Sharma, S. J. Halme, M. M. Butusov: *Optical Fiber Systems and Their Components*, Springer Ser. Opt. Sci., Vol. 24 (Springer, Berlin, Heidelberg 1981)

Chapter 5

5.1 V. A. Fabricant: Tr. Vses. Elektrotekh. Inst. **41**, 254 (1940)
5.2 J. P. Gordon, H. J. Zeiger, C. H. Townes: Phys. Rev. **95**, 282 (1954) and **99**, 1264 (1955)
5.3 K. Shimoda: In *Proc. Symposium on Optical Masers*, ed. by J. Fox (Polytechnic Press, New York 1963) p. 95
5.4 M. Maeda, N. B. Abraham: Phys. Rev. A**26**, 3395 (1982);
C. O. Weiss, H. King: Opt. Commun. **44**, 59 (1982);
R. Hauck, F. Hollinger, H. Weber: Opt. Commun. **47**, 141 (1983);
L. A. Lugiato, L. M. Narducci, D. K. Bandy, C. A. Pennise: Opt. Commun. **46**, 64 (1983)

Chapter 6

6.1 K. Shimoda: In *Proc. Symposium on Optical Masers*, ed. by J. Fox (Polytechnic Press, New York 1963) p. 95

Chapter 7

7.1 R. P. Feynman, F. L. Vernon, Jr., R. W. Hellworth: J. Appl. Phys. **28**, 49 (1957)
7.2 R. G. DeVoe, R. G. Brewer: Phys. Rev. Lett. **50**, 1269 (1983)
7.3 T. Endo, T. Muramoto, T. Hashi: Opt. Commun. **51**, 163 (1984)

Chapter 8

8.1 I. I. Sobleman, L. A. Vainshtein, E. A. Yukov: *Excitation of Atoms and Broadening of Spectral Lines*, Springer Ser. Chem. Phys., Vol. 7 (Springer, Berlin, Heidelberg 1981)
8.2 D. C. Hanna, M. A. Yuratich, D. Cotter: *Nonlinear Optics of Free Atoms and Molecules*, Springer Ser. Opt. Sci., Vol. 17 (Springer, Berlin, Heidelberg 1979)
8.3 K. Shimoda (ed.): *High-Resolution Laser Spectroscopy*, Topics Appl. Phys., Vol. 13 (Springer, Berlin, Heidelberg 1976)
8.4 M. S. Feld, V. S. Letokhov (eds.): *Coherent Nonlinear Optics*, Topics Curr. Phys., Vol. 21 (Springer, Berlin, Heidelberg 1980)
8.5 J. H. Eberly, L. Allen: *Optical Resonance and Two-Level Atoms* (Wiley, New York 1974)
8.6 C. P. Slichter: *Principles of Magnetic Resonance*, 2nd ed., Springer Ser. Solid-State Sci., Vol. 1 (Springer, Berlin, Heidelberg 1980)

Chapter 9

9.1 K. Shimoda, T. C. Wang, C. H. Townes: Phys. Rev. **102**, 1308 (1956)
9.2 B. van der Pol: Proc. IRE **22**, 1051 (1934)
9.3 K. Shimoda, T. Yajima, Y. Ueda, T. Shimizu, T. Kasuya: *Quantum Electronics I* (Shokabo, Tokyo 1972) [in Japanese]
9.4 M. Sargent III, M. O. Scully, W. E. Lamb, Jr.: *Laser Physics* (Addison-Wesley, Reading 1974)
9.5 S. E. Harris, O. P. McDuff: IEEE J. QE-**1**, 245 (1965)
9.6 W. E. Lamb, Jr.: Phys. Rev. **134**, A1429 (1964)
9.7 B. L. Gyorffy, M. Borenstein, W. E. Lamb, Jr.: Phys. Rev. **169**, 340 (1968)
9.8 S. Stenholm, W. E. Lamb, Jr.: Phys. Rev. **181**, 618 (1969)
9.9 B. J. Feldman, M. S. Feld: Phys. Rev. A**1**, 1375 (1970)
9.10 K. Shimoda, K. Uehara: Jpn. J. Appl. Phys. **10**, 460 (1971)
9.11 K. Uehara, K. Shimoda: Jpn. J. Appl. Phys. **10**, 623 (1971)
9.12 H. Haken: *Laser Theory* (Springer, Berlin, Heidelberg 1984)
9.13 D. Marcuse: *Principles of Quantum Electronics* (Academic, New York 1980)
9.14 H. Risken, C. Schmid, W. Weidlich: Z. Phys. **193**, 37 (1966)
9.15 M. Lax, W. H. Louisell: IEEE J. QE-**3**, 47 (1967)

Subject Index

H. Haken

Laser Theory

Corrected printing. 1984. 72 figures. XV, 320 pages. ISBN 3-540-12188-9 (Originally published as "Handbuch der Physik/Encyclopedia of Physics, Volume 25, 2c", 1970)

Contents: Introduction. – Optical resonators. – Quantum mechanical equations of the light field and the atoms without losses. – Dissipation and fluctuation of quantum systems. The realistic laser equations. – Properties of quantized electromagnetic fields. – Fully quantum mechanical solutions of the laser equations. – The semiclassical approach and its applications. – Rate equations and their applications. – Further methods for dealing with quantum systems far from thermal equilibrium. – Appendix: Useful operator techniques. – Sachverzeichnis (Deutsch-Englisch). – Subject Index (English-German).

W. Koechner

Solid-State Laser Engineering

1976. 287 figures, 38 tables. XI, 620 pages. (Springer Series in Optical Sciences, Volume 1). ISBN 3-540-90167-1

"My reaction to this book is that I wish that it had been available years ago when solid-state lasers were originally being designed. It is a very practical book containing a wealth of detail about how solid-state lasers are actually constructed. It describes how many of the problems involved in their construction and operation have been overcome. There is sufficient theoretical and mathematical background to understand the operation of the devices and to make the book self-contained ..."

J. Opt. Soc. Am.

M. Young

Optics and Lasers

Including Fibers and Integrated Optics

2nd revised edition. 1984. 160 figures. XIX, 269 pages. (Springer Series in Optical Sciences, Volume 5). ISBN 3-540-13014-4

"... The book is really delightful optics reading, unusually wide ranging, up-to-date, concise, and useful ... Examples of some up-to-date topics included are optical processing, MTF, acousto-optical modulation, optical waveguides, Fourier transform spectroscopy, and holography. Each subject is treated stressing physical principles, and seldom is a subject left without at least one constructive, non-obvious comment ... it has great value for the practicing physicist or engineer who works from time to time with optics. Moreover the material is certainly appropriate to an introductory undergraduate course in optics ... The book also makes an excellent review of optics for the graduate student or applied scientist who wants to rethink things through ..."

J. Opt. Soc. America

Excimer Lasers

Editor: **C. K. Rhodes**

2nd enlarged edition. 1984. 100 figures. XII, 271 pages. (Topics in Applied Physics, Volume 30). ISBN 3-540-13013-6

Contents: *P. W. Hoff, C. K. Rhodes:* Introduction. – *M. Krauss, F. H. Mies:* Electronic Structure and Radiative Transitions of Excimer Systems. – *M. V. McCusker:* The Rare Gas Excimers. – *C. A. Brau:* Rare Gas Halogen Excimers. – *A. Gallagher:* Metal Vapor Excimers. – *D. L. Huestis, G. Marowsky, F. K. Tittel:* Triatomic Rare-Gas-Halide Excimers. – *H. Pummer, H. Egger, C. K. Rhodes:* High-Spectral-Brightness Excimer Systems. – *K. Hohla, H. Pummer, C. K. Rhodes:* Applications of Excimer Systems. – List of Figures. – List of Tables. – Subject Index.

Springer-Verlag Berlin Heidelberg New York Tokyo

W. Demtröder

Laser Spectroscopy

Basic Concepts and Instrumentation
2nd corrected printing. 1982. 431 figures. XIII, 696 pages. (Springer Series in Chemical Physics, Volume 5). ISBN 3-540-10343-0

From the reviews: "The scope of this book is most impressive. It is authoritative, illuminating and up-to-date. The 650 pages of text are supplemented by 34 pages of references, and many of the chapters are furnished with a selection of problems. It is strongly recommended for all spectroscopists of the laser era and will be valuable for research students entering spectroscopic laboratories."

Contemporary Physics

S. A. Losev

Gasdynamic Laser

1981. 100 figures. X, 297 pages. (Springer Series in Chemical Physics, Volume 12). ISBN 3-540-10503-4

Contents: Introduction. - Basic Concepts of Quantum Electronics. - Physico-Chemical Gas Kinetics. - Relaxation in Nozzle Gas Flow. - Infrared CO_2 Gasdynamic Laser. - Gasdynamic Lasers with Other Active Medium. - Appendixes. - List of the Most Used Symbols. - References. - Subject Index.

G. Brederlow, E. Fill, K. J. Witte

The High-Power Iodine Laser

1983. 46 figures. IX, 182 pages. (Springer Series in Optical Sciences, Volume 34). ISBN 3-540-11792-X

Contents: Introduction. - Basic Features. - Principles of High-Power Operation. - Beam Quality and Losses. - Design and Layout of an Iodine Laser System. - The ASTERIX III System. - Scalability and Prospect of the Iodine Laser. - Conclusion. - References. - Subject Index.

Springer-Verlag
Berlin
Heidelberg
New York
Tokyo